Security Sound Bites

Important Ideas About Security From
Smart-Ass, Dumb-Ass, and Kick-Ass Quotations

Roger G. Johnston

To Janie.

Table of Contents

1 - Introduction

I've given a lot of talks about physical security and vulnerability assessments. For many years, I would include a short, entertaining quote at the bottom of some of my slides to emphasize certain points I was trying to make. I wouldn't comment on the quotes directly, just let the audience read them on their own if they were interested. People seemed to like this approach.

At first, I chose quotes with rather straightforward connections to the material I was presenting. But then I discovered something interesting. If the quote was only tangentially related to the point I was trying to make, people afterwards would get into quite lively discussions about the meaning of the quote. In the process, they were thinking more profoundly about the security issues I was trying to raise than would happen otherwise.

This book contains many of the quotes I have used over the years, as well as various observations, maxims, and anecdotes. A lot of these are clearly about security. Others aren't exactly about security, but maybe they should be. Or maybe they really are about security if you look at them in the right way.

Many of these items are humorous. Others are thought provoking. Some are thought provoking and humorous. A lot are pithy or amazing or cynical. Many are remarkable in their observational clarity or elegance. More than a few are just plain dumb or disturbing. All together, they capture fundamental and profound truths about security that can be ignored only at great peril.

I hope you find these various sound bites entertaining. If so...great, but that's not the point. If you think carefully about the ideas that each of these snippets represents, then take the lessons they offer to heart, I guarantee you'll have a better security program. If nothing else, sometimes humor, elegance, cynicism, or boldface stupidity can kick us out of our day-to-day mental rut and get us thinking in new directions.

These quotations can be used as a springboard for discussion with your security colleagues, or as an interesting way to open a meeting or close an email. Or they can be something to ponder on your own.

At the start of each section is a list of some of the key points that I think the sound bites in that section make—or at least hint at. My suggestion is that you read the quotations in each section with these key points in mind. You may well find other, better meanings, too, that I have not paraphrased.

Hopefully in the process of considering these snippets, you can come to some fresh insights about security, security management, organizational behavior, or even just the human condition. These things are, of course, intricately connected. Fundamentally, security is about human beings and all their faults, foibles, and flaws, not to mention their idiosyncrasies, foolishness, maliciousness, arrogance, ignorance, genius, wit, wisdom, courage, stupidity, vision, self-delusion, self-sacrifice, and lunacy. All of which can be found reflected in these sound bites.

The views expressed here are my own and should not necessarily be attributed to my employer, the people quoted here, or anybody else, sane or otherwise. It should also not be assumed that I agree with the words or ideas contained or implied in any given quotation or anecdote that appears in this book. I use some quotations as horrible examples of thinking gone dreadfully wrong, while others I hold up as shining examples of enlightened reasoning.

To the best of my knowledge, all items included here fall under the fair use or public domain guidelines of copyright law in the United States. Quotations remain the intellectual property of their respective originators. I make no claim of copyright for individual quotations, observations, or anecdotes that I did not originate. By quoting any given person here, I do not mean to imply that they have endorsed or approved this book.

I make no claims for the accuracy of the quotes or who gets credit for them. Many quotes are attributed to "Anonymous", but it is likely that at least their mom knew them. If you're the originator of the quote, my apologies for not giving you the credit. If I've given credit to the wrong person, again my apologies.

Quotations tend to naturally get tweaked over time to both sound better and reinforce their message. Regarding the accuracy of the quotes in this book, I take solace in the following quotes:

Famous remarks are very seldom quoted correctly -- Simeon Strunsky (1879-1948)

Quotation (n): The act of repeating erroneously the words of another. -- Ambrose Bierce (1842-1914?)

Introduction

What's the use of a good quotation if you can't change it? -- Doctor Who

I have the hubris to quote myself on occasion when I can't find quotes from others to make the desired points. The intent, however, is not to try to place myself either at the lofty heights of some of the brilliant minds I quote (e.g., Albert Einstein, Demosthenes, Yogi Berra), nor at the frightening depths of the flaming morons who also appear.

Finally, thanks Dr. Jon Warner for recommending some quotes.

2 - Cyber Security

Some Key Points:

• **It's more about the user than the security staff or technology.**

• **Effective IT security requires effective physical security.**

There are only two industries that refer to their customers as "users".
-- Edward Tufte

They have computers, and they may have other weapons of mass destruction.
-- Attorney General Janet Reno

Even the smartest IT security staff is no match for user ignorance.
-- Michael Perry

The methods that most effectively minimize the ability of intruders to compromise information security are comprehensive user training and education. Enacting policies and procedures simply won't suffice. Even with oversight, the policies and procedures may not be effective.
-- Kevin Mitnick

Without physical security, no other security measures can be considered effective.
-- Tom Caddy

It does little good to have great computer security if wiring closets are easily accessible or individuals can readily walk into an office and sit down at a computer and gain access to systems and applications. Even though the skill level required to hack systems and write viruses is becoming widespread, the skill required to wield an ax, hammer, or fire hose and do thousands of dollars in damage is even more widely held.
-- Michael Erbschloe

Computer Science is no more about computers than Astronomy is about telescopes.
-- E. W. Dijkstra

A Bus Station is where buses stop. A Train Station is where trains stop. On my desk, there is a Work Station.
    -- Jojn Wätte

If you don't know how to do something, you don't know how to do it with a computer.
    -- Anonymous

Computers make it easier to do a lot of things, but most of the things they make it easier to do don't need to be done.
    -- Andy Rooney

Is it the computer's fault for freezing, or our fault for trusting the worthless piece of crap to begin with?
    -- Anonymous

The user's going to pick dancing pigs over security every time.
    -- Bruce Schneier

Actual news story: On February 13, 2009, InfoMedia, Inc., which developed iFart for the iPhone, filed a lawsuit in federal district court in Colorado against Air-O-Matic, Inc., maker of the competing app "Pull My Finger" over trademark rights to the phrase "pull my finger". Both apps simulate farting noises. There are at least 75 total flatulence simulation apps available. [*Author's Comment*: It's pretty darn clear that this kind of software is exactly what *Rear* Admiral Grace Hopper had in mind when she was pioneering the development of software.]

Definition—portable: (adjective)-exposed to a mutable ownership through vicissitudes of possession.
    -- Anonymous

Compaq was considering changing the command "Press Any Key" to "Press Enter Key" because of the flood of calls asking where the Any Key is.

How long is this Beta guy going to keep testing our stuff?"
    -- Inquiry from a senior manager

Who were the beta testers for Preparations A through G?
    -- Bumper Sticker

Any Internet user knows it is quite difficult to stumble across pornography.

-- Sen. Russell Feingold

I don't understand computers. I don't even understand people who understand computers.
    -- Queen Juliana of the Netherlands

Percentage of Canadians who say they approve of the information superhighway: 63%. Percentage of Canadians who say they know what the information superhighway is: 54%
    -- PC Magazine

A computer once beat me at chess, but it was no match for me at kick boxing.
    -- Emo Philips

On two occasions I have been asked by members of Parliament, "Pray, Mr. Babbage, if you put into the machine wrong figures, will the right answers come out?" I am not able rightly to apprehend the kind of confusion of ideas that could provoke such a question.
    -- Charles Babbage (1791-1871)

If you put tomfoolery into a computer, nothing comes out of it but tomfoolery. But this tomfoolery, having passed through a very expensive machine, is somehow ennobled and no-one dares criticize it.
    -- Pierre Gallois

Computers are useless. They can only give you answers.
    -- Pablo Picasso (1881–1973)

I just feel so sad for the human race after using a PC.
    -- Anonymous Mac user

Actual call to a computer help line:
Customer: I bought your fancy graphics card, and my Windows display isn't any better than it was before.
Tech support guy: We'd better look at the installation then.
Customer: You mean I have to install it?

The only truly secure system is one that is powered off, cast in a block of concrete, and sealed in a lead-lined room with armed guards—and even then I have my doubts.
    -- Gene Spafford

The only system which is truly secure is one which is switched off and unplugged, locked in a titanium-lined safe, buried in a concrete bunker, and is surrounded by nerve gas and very highly paid armed guards. Even then, I wouldn't stake my life on it.
    -- Gene Spafford

If you spend more on coffee than on IT security, you will be hacked. What's more, you deserve to be hacked.
    -- Richard Clarke, White House Cyber Security Advisor

The first phone call to Michael Wolff's NetGuide phone hotline, January 2, 1994: "Hello! Is this the Internet?"

Actual Tech Support phone conversation:
Customer: I can't get on the Internet.
Tech support: Are you sure you used the right password?
Customer: Yes, I'm sure. I saw my colleague do it.
Tech support: Can you tell me what the password was?
Customer: Five stars.

Actual call to a computer tech support line: I have a huge problem. A friend of mine has placed a screen saver on my computer, but every time I move the mouse, it disappears.

Factoid: Microsoft Office was first developed for the Mac, in 1989. A Windows version came out in 1990.

Kernighan's Law: Debugging is twice as hard as writing the code in the first place. Therefore, if you write the code as cleverly as possible, you are, by definition, not smart enough to debug it.

I conclude that there are two ways of constructing a software design: One way is to make it so simple that there are obviously no deficiencies and the other way is to make it so complicated that there are no obvious deficiencies.
    -- Charles Hoare

A Dell technician received a call from a customer who was enraged because his computer had told him he was "bad and an invalid".
    -- Wall Street Journal

Cyber Security

Actual Call:

Tech support: Good day. How may I help you?

Male customer: Hello... I can't print.

Tech support: Would you click on "start" for me and...

Customer: Listen pal! Don't start getting technical on me! I'm not Bill Gates.

You can hardly tell where the computer models finish and the real dinosaurs begin.
-- Laura Dern, actress in *Jurassic Park* (1993)

3 - Ciphers & Cryptography

Some Key Points:

• **Ciphers do not offer absolute security.**

• **Ciphers and Data Authentication have their uses, but they add little to zero extra security if you lack good physical security, a good security culture, and haven't dealt adequately with the insider threat.**

• **They don't legitimize raw data that you can't believe in the first place, or data collected, stored, generated, or transmitted by hardware or software that is not secure, or handled by untrustworthy personnel at either end.**

Definition—encryption: (1) Attempting to secure the communications channel between two hopelessly unsecure locations or devices, each controlled, designed, or manufactured by completely untrustworthy or incompetent personnel. (2) A magic band-aid that fixes all security flaws, even those having nothing to do with data or communications security. (3) Evidence that nobody has spent any time thinking about security.

Definition—data authentication: A magical technique that gives us 100% confidence in the veracity of data even though the machine that generated or transmitted it, and the people who handle or made the machine can't be trusted, are knuckleheads, and the data is probably wrong anyway.

Never underestimate the time, expense, and effort an opponent will expend to break a code.
    -- Robert Morris

Using encryption on the Internet is the equivalent of arranging an armored car to deliver credit-card information from someone living in a cardboard box to someone living on a park bench.
    -- Gene Spafford

There is no assurance that a foreign government cannot also "break" the [DSS] system, running the risk of a "digital Pearl Harbor"—a devastating loss of the security of the entire national financial and business transaction systems.

-- D. James Bidzos

Definition—computationally secure: (adjective)-A weasel term applied to the security of ciphers that really means, "We're not imaginative enough to envision a successful attack."

Factoid: The one-time keypad (Vernam cipher) is the only cipher that can be shown mathematically to be unbreakable.

When cryptography is outlawed, bayl bhgynjf jvyy unir cevinpl
    -- 6&nL#bi~8r!

One of the most singular characteristics of the art of deciphering is the strong conviction possessed by every person, even moderately acquainted with it, that he is able to construct a cipher which nobody else can decipher. I have also observed that the cleverer the person, the more intimate is his conviction. In my earliest study of the subject, I shared in this belief, and maintained it for many years.
    -- Charles Babbage (1791-1871)

The security of a cipher lies less with the cleverness of the inventor than with the stupidity of the men who are using it.
    -- Waldemar Werther

Anyone who attempts to generate random numbers by deterministic means is, of course, living in a state of sin.
    -- John von Neumann (1903-1957)

Red Herring Maxim: At some point in any challenging security application, somebody (or nearly everybody) will propose or deploy more or less pointless encryption, hashes, or data authentication along with the often incorrect and largely irrelevant statement that "the cipher [or hash or authentication algorithm] cannot be broken".

Comments: Product anti-counterfeiting tags and International Nuclear Safeguards are two security applications highly susceptible to this fuzzy thinking.

With product anti-counterfeiting tags, it is no harder for the product counterfeiters to make copies of encrypted data than it is to make copies of unencrypted data. They don't have to understand the encryption scheme or the encrypted data to copy it, so that the degree of difficulty in breaking the encryption (usually overstated) is irrelevant. Indeed, if there was a technology that could preventing cloning of encrypted data (or hashes or digital authentication), then that same technology could be used to prevent cloning of the unencrypted original data, in which case the encryption has no significant role to play. (Sometimes one might wish to send secure information to counterfeit hunters in the field, but the security features and encryption typically employed on cell phones or computers is good enough. It is not necessary to pay for expensive, proprietary encryption—which may well be quite weak.)

What makes no sense is putting encrypted data on a product, with or without it including encrypted data about an attached anti-counterfeiting tag; the bad guys can easily clone the encrypted data without having to understand it. When there is an anti-counterfeiting tag on a product, only the degree of difficulty of cloning it is relevant, not the encryption scheme. The use of unique, one-of-a-kind tags (i.e., complexity tags) does not alter the relative unimportance of the encryption as an anti-counterfeiting measure.

Sometimes people promoting encryption for product anti-counterfeiting vaguely have in mind an overly complicated (and usually incomplete/flawed) form of a virtual numeric token ("call-back strategy"). ([See RG Johnston, "An Anti-Counterfeiting Strategy Using Numeric Tokens", *International Journal of Pharmaceutical Medicine* 19, 163-171 (2005).] Or they may confuse "serialization" with virtual numeric tokens.

Encryption is also often thought of as a silver bullet for International Nuclear Safeguards. The fact is that encryption or data authentication is of little security value if the adversary can easily break into the equipment holding the secret key without detection (as is usually the case), if there is a serious insider threat that puts the secret encryption key at risk (which is pretty much always the case), and/or if the surveillance or monitoring equipment containing the secret key is designed, controlled, inspected, maintained, stored, observed, or operated by the adversary (as is typically the case in International Nuclear Safeguards).

4 - Psychology

Some Key Points:

• **People are complex, flawed, and weird.**

• **You can't understand security without understanding psychology.**

• **Humor is a powerful tool for leaders, to fight the negative effects of cognitive dissonance, and to encourage creative thinking and problem solving.**

Humor is a good test for sanity. If you laugh, you are sane. Unless, of course, you laugh constantly, at nothing at all.
 -- Anonymous

To crack a serious problem, crack a joke.
 -- Haresh Sippy

The people who fear humor, and they are many, are suspicious of its power to present things in unexpected lights, to question received opinions, and to suggest unforeseen possibilities.
 -- Robertson Davies (1913-1995)

I dream of a better tomorrow, where chickens can cross the road and not be questioned about their motives.
 -- Ralph Waldo Emerson (1803-1882)

A person reveals his character by nothing so clearly as the joke he resents.
 -- Georg C. Lichtenberg (1742-1799)

You grow up the day you have your first real laugh—at yourself.
 -- Ethel Barrymore (1879-1959)

Mix a little foolishness with your prudence: It's good to be silly at the right moment.
 -- Horace (65 – 8 BC)

Comedy is simply a funny way of being serious.
> -- Peter Ustinov (1921-2004)

Anything worth taking seriously is worth making fun of.
> -- Anonymous

The human race has one really effective weapon, and that is laughter.
> -- Mark Twain (1835-1910)

Most men are within a finger's breadth of being mad.
> -- Diogenes the Cynic (412–323 BC)

The only normal people are the ones you don't know very well.
> -- Joe Anci

Remember as far as anyone knows, we're a nice normal family.
> -- Homer Simpson

Normal is not something to aspire to, it's something to get away from.
> -- Actress and Director Jodie Foster

You know what 'Normal' is? A setting on a washing machine. No one wants to be that.
> -- Ashley Purdy

Don't accept rides from strange men, and remember that all men are strange as hell.
> -- Robin Morgan

People will believe anything if you whisper it.
> -- Anonymous

Why do we press harder on the TV remote keys when the batteries are getting weak?
> -- Anonymous

That which does not kill us, only makes us stranger.
> -- Aeon Flux

If the brain were so simple we could understand it, we would be so simple we couldn't.
> -- Lyall Watson (1939-2008)

Psychology

It is easier to understand Man in general than to understand one man in particular.
  -- François de La Rochefoucauld (1613-1680)

You talk to God, you're religious. God talks to you, you're psychotic.
  -- Doris Egan

A man generally has two reasons for doing a thing. One that sounds good, and a real one.
  -- J. Pierpoint Morgan (1837-1913)

When truth is discovered by someone else, it loses something of its attractiveness.
  -- Alexander Solzhenitsyn (1918-2008)

He was a great patriot, a humanitarian, a loyal friend; provided, of course, he really is dead.
  -- Voltaire (1694-1778)

Anyone who goes to a psychiatrist ought to have his head examined.
  -- Samuel Goldwyn (1879-1974)

I told my psychiatrist that everybody hates me. He said I was being ridiculous—everybody hasn't met me yet.
  -- Rodney Dangerfield (1921-1997)

There are many kinds of intelligence. Scientists just haven't identified mine yet.
  -- Anonymous

Understanding is reached only after confrontation.
  -- Miss Ivannah, the topless fortune teller in *Mallrats* (1995)

To predict the behavior of ordinary people in advance, you only have to assume that they will always try to escape a disagreeable situation with the smallest possible expenditure of intelligence.
  -- Friedrich Nietzsche (1844-1900)

The public will believe anything, so long as it is not founded on truth.
  -- Edith Sitwell (1887-1964)

Nobody eats at that restaurant anymore because it's always so crowded.
  -- Yogi Berra

Factoid: Sports teams that wear dark uniforms are penalized more than teams that wear white uniforms.

Man is the only animal that laughs and weeps; for he is the only animal that is struck with the difference between what things are and what they ought to be.
-- William Hazitt (1778-1830)

A harmless hilarity and a buoyant cheerfulness are not infrequent concomitants of genius; and we are never more deceived than when we mistake gravity for greatness, solemnity for science, and pomposity for erudition.
-- C.C. Colton (1780-1832)

It is our responsibilities, not ourselves, that we should take seriously.
-- Peter Ustinov (1921-2004)

We must avoid here two complementary errors: on the one hand that the world has a unique, intrinsic, pre-existing structure awaiting our grasp; and on the other hand that the world is in utter chaos. The first error is that of the student who marvelled at how the astronomers could find out the true names of distant constellations. The second error is that of the Lewis Carroll's Walrus who grouped shoes with ships and sealing wax, and cabbages with kings...
-- R. Abel

If trees could scream, would we be so cavalier about cutting them down? We might, if they screamed all the time, for no good reason.
-- Jack Handey

That's the difference between me and the rest of the world! Happiness isn't good enough for me! I demand euphoria!
-- Bill Watterson, from *Calvin and Hobbes*

The capriciousness of our temperament is even stranger than the whims of fortune.
-- François de La Rochefoucauld (1613-1680)

This guy walks into a psychiatrist's office and says, "Doc, you've got to help me!" "Sure," says the psychiatrist, "What's the problem?" "Well, Doc, it's my brother. He thinks he's a chicken." "My goodness!" says the psychiatrist, "How long has this been going on?" "About seven years," says the man. "Seven years!" exclaims the psychiatrist. "Why didn't you come to me sooner?" "Well I would've," says the man, "but we needed the eggs."

-- old Vaudeville joke

We are more ready to try the untried when what we do is inconsequential. Hence the fact that many inventions had their birth as toys.
-- Eric Hoffer (1902-1983)

To punish me for my contempt for authority, fate made me an authority myself.
-- Albert Einstein (1879-1955)

Don't tell me that worry doesn't do any good. I know better! The things I worry about don't happen!
-- Anonymous

If you think you can, or you think you can't, you are right.
-- Henry Ford (1863-1947)

The food here is terrible, and the portions are so small.
-- Woody Allen

If you want to know what God thinks of money, look at the people he gave it to.
-- Dorothy Parker (1893-1967)

In my opinion, we don't devote nearly enough scientific research to finding a cure for jerks.
-- Bill Watterson, from *Calvin and Hobbes*

5 - Polygraphs

Some Key Points:

• **Current polygraphs (lie detectors) are pseudo-scientific nonsense.**

The polygraph is a ruse, carefully constructed as a tool of intimidation, and used as an excuse to conduct illegal inquisition under psychologically and physically unpleasant circumstances.
-- The Committee for the Scientific Investigations of Claims of the Paranormal (CSICOP)

Factoid: No persons captured as U.S. spies in the last 3 decades failed a polygraph exam. A number passed polygraph exams multiple times.

In 2002, the National Academy of Sciences completed an independent, $860,000 study on the effectiveness of polygraphs ("lie detectors"). See http://www.nap.edu/books/0309084369/html.
Some conclusions from this study:
• "Polygraph test accuracy may be degraded by countermeasures…"
• "…overconfidence in the polygraph—a belief in its accuracy that goes beyond what is justified by the evidence—…presents a danger to national security…"
• "Its accuracy in distinguishing actual or potential security violators from innocent test takers is insufficient to justify reliance on its use in employee security screening…"

Definition—polygraph, a.k.a. "lie detector": A pseudo-scientific device invented by William Marston in the 1920's with about as much grounding in reality as his other major invention (the comic book character Wonder Woman).

A polygraph does not detect lies. It detects physiological responses which are not well correlated with dishonesty. If you're a narcissist, or you believe your own lies, or you not particularly emotionally responsive, or you know the various techniques to fool the polygraph, or the polygraph examiner likes you, you won't fail the "exam" even if you don't tell the truth.
-- Anonymous

Polygraphs

You don't fool a polygraph, you fool the polygraph examiner.
-- Anonymous

Factoid: It has been pointed out that most states require more training to become a licensed barber than to become a certified polygraph examiner.

6 - Security Culture

Some Key Points:

• "Security Culture" is the official and unofficial, formal and informal behaviors, attitudes, perceptions, strategies, rules, policies, and practices associated with security. An organization is unlikely to have good security without a good security culture. Sometimes, the security hardware is considered a component of security culture (in the same sense that archaeologists consider physical artifacts to be part of culture), but this isn't especially helpful. Sometimes the unofficial and informal aspects are separately called "security climate".

• Your security is no better than your security culture and climate.

• People are the security.

• Nobody can foresee all the threats and vulnerabilities, and it's silly and ignorant to expect them to.

• A good security culture needs to be based on motivating employees, not threatening them.

• Like the weather, everybody knows that Security Culture is important, but nobody does anything about it!

Be Afraid, Be Very Afraid Maxim: If you're not running scared, you have bad security or a bad security product.

The best safety lies in fear.
    -- William Shakespeare (1564-1616), *Hamlet*, 1:3

Health is not simply the absence of sickness.
    -- Hannah Green

<u>Accountability Maxim</u>: Organizations that talk a lot about holding people accountable for security will never have good security. Security needs to be motivated, not threatened.

Firing people does not engender accountability, just cover-ups, scapegoating, and deceit. It also makes security the enemy of employees.
    -- Anonymous

Distrust all in whom the impulse to punish is powerful.
    -- Friedrich Nietzsche (1844-1900)

<u>Scapegoat Maxim</u>: The main purpose of an official inquiry after a serious security incident is to find somebody to blame, not to fix the problems.

When all candles be out, all cats be grey.
    -- John Heywood, (1497-1580)

There is nothing more arrogant than a security officer.
    -- Richard Johnson

What's Wrong with This Picture?
"I think the worst problem was the way the security was set up for this particular project. The people who set it up were actually trying to be very conscious of security, but they didn't make a plan that addressed all the potential risks."
    -- Testimony to Congress after yet another serious security incident at Los Alamos National Laboratory

<u>Somebody Must've Thought It Through Maxim</u>: The more important the security application, the less careful and critical thought and analysis has gone into it.

Those security guys are really starting to get on my nerves.
    -- From the movie, *Menno's Mind* (1996)

Protection and security are only valuable if they do not cramp life excessively.
    -- Carl Jung (1875-1961)

Jack Byrnes: Trust me, Greg. When you start having little Fockers running around, you'll feel the need for this type of security.
    -- From the movie, *Meet the Parents* (2000)

Don't confuse control with security.

-- Bruce Schneier

The difference between an American and a European is that a European thinks that 100 miles is a long distance while an American thinks that 100 years is a long time.
    -- Anonymous

7 - The Insider Threat & Social Engineering

Some Key Points:

• **Acknowledging the insider threat is a challenge.**

• **Mitigating it is even harder**

• **Much of the insider threat is non-deliberate (complacent and careless employees), but still dangerous.**

• **Troublemakers aren't automatically an insider threat.**

• **Employee perceptions of fairness, not objective reality, are all that matter for employee disgruntlement—a key factor for insider threat.**

• **People who are treated badly but expect to be don't tend to be disgruntled.**

• **People who appear to be treated well may be disgruntled over seemingly minor issues.**

• **Background checks rely primarily on information provided by the subject.**

We have met the enemy and he is us.
  -- Walt Kelly (1913-1973), the words of Pogo in an Earth Day 1971 cartoon strip

We Have Met the Enemy and He is Us Maxim: The insider threat from careless or complacent employees and contractors exceeds the threat from malicious insiders (though the latter is not negligible.)

Never marry a man you wouldn't want to be divorced from.
  -- Nora Ephron (1941-2012)

No one can build his security upon the nobleness of another person.
  -- Willa Cather (1873-1947)

The enemy is anybody who's going to get you killed, no matter which side he's on.

-- Joseph Heller (1923-1999), *Catch-22*

Some quit and leave. Others quit and stay.
    – Anonymous

Everybody has a price. If not, everybody has a weakness.
    -- Michelle Bulleri

An autobiography is only to be trusted when it reveals something disgraceful.
    -- George Orwell (1903-1950)

Actual courtroom testimony:
Q: You said you went to Galveston in 1920, yet the first job you told me about was in 1946. What did you do between 1920 and 1946?
A: Well, I didn't go to work as soon as I got there.

People often represent the weakest link in the security chain and are chronically responsible for the failure of security systems.
    -- Bruce Schneier

Insider Risk Maxim: Most organizations will ignore or seriously underestimate the threat from insiders.

There are some who become spies for money, or out of vanity and megalomania, or out of ambition, or out of a desire for thrills. But the malady of our time is of those who become spies out of idealism.
    -- Max Lerner (1902-1992)

The desire to reveal is greater than the desire to conceal.
    -- Carl Jung (1875-1961)

Honesty may be the best policy, but it's important to remember that apparently, by elimination, dishonesty is the second-best policy.
    -- George Carlin (1937-2008)

Whatever you condemn, you have done yourself.
    -- Georg Groddeck (1866-1934)

Show me a liar, and I will show thee a thief.
    -- George Edward Herbert (1866-1923)

Whoever washes the dishes is the one who breaks them.
    -- old English proverb

There is no safety for honest men but by believing all possible evil of evil men.
    -- Edmund Burke (1729-1797)

Who naught suspects is easily deceived.
    -- Francesco Petrarch (1304-1374)

It takes greater virtues to withstand good fortune than bad fortune.
    -- François de La Rochefoucauld (1613-1680)

I have not a particle of confidence in a man who has no redeeming vices.
    -- Mark Twain (1835-1910)

It has been my experience that folks who have no vices have very few virtues.
    -- Abraham Lincoln (1809-1865)

It's awful hard to get people interested in corruption unless they can get some of it.
    -- Will Rogers (1879-1935)

Envy aims very high.
    -- Ovid (43 BC–17 AD)

The true hypocrite is the one who ceases to perceive his deception, the one who lies with sincerity.
    -- André Gide

Put more trust in nobility of character than in an oath.
    -- Solon (~630 - ~560 BC)

Everybody is a moon, and has a dark side which he never shows to anybody.
    -- Mark Twain (1835-1910)

There's a deception to every rule.
    -- Hal Lee Luyah

A man isn't honest simply because he's never had a chance to steal.
    -- Yiddish proverb

Many are saved from sin by being so inept at it.

-- Mignon McLaughlin (1913-1983)

Slight are the outward signs of evil thought.
   -- old proverb

Those you trust the most can steal the most.
   -- David Pauly

A thing worth having is a thing worth cheating for.
   -- W.C. Fields (1880-1946)

Money never made a man happy yet, or will it. The more a man has, the more he wants. Instead of filling a vacuum, it makes one.
   -- Ben Franklin (1706-1790)

I'm not corrupt, I'm morally flexible.
   -- Anonymous

Evil: That which one believes of others.
   -- H.L. Mencken (1880-1956)

Why do we never expect dull people to be rascals?
   -- Mason Cooley (1927-2002)

-But I'm telling you, I'm an anti-Communist!
-I don't care what kind of Communist you are, get outta here!
   -- Exchange between Harry Warner and a Warner Brothers executive during the
      Red Scare

I worry that the person who thought up Muzak may be thinking up something else.
   -- Lily Tomlin

Factoid: No captured major U.S. spy was mentally ill at the time of his capture. They were jerks, traitors, and narcissists, certainly, but not crazy.

Don't place too much confidence in the man who boasts of being as honest as the day is long. Wait until you meet him at night.
   -- Robert C. Edwards

If only there were evil people somewhere insidiously committing evil deeds, and it were necessary only to separate them from the rest of us and destroy them. But the line dividing good and evil cuts through the heart of every human being, and who is willing to destroy his own heart?
-- Alexander Solzhenitsyn (1918-2008)

The world is a stage, but the play is badly cast.
-- Oscar Wilde (1854-1900)

Factoid: Out of 30 million known species of bacteria, only about 70 cause disease.

Honest and sincere acts mislead the wicked and cause them to lose their path to their own goals, because mean-spirited people usually believe that people never act without deceit.
-- Madame de Sablé (1599-1678)

Nothing can tell us so much about the general lawlessness of humanity as a perfect acquaintance with our own immoderate behavior. If we would think over our own impulses, we would recognize in our own souls the guiding principle of all vices which we reproach in other people; and if it is not in our very actions, it will be present at least in our impulses.
-- Madame de Sablé (1599-1678)

When we see men of a contrary character, we should turn inwards and examine ourselves.
-- Confucius (551 – 479 BC)

Many a man's reputation would not know his character if they met on the street.
-- Elbert Hubbard (1856-1915)

Many of us believe that wrongs aren't wrong if it's done by nice people like ourselves.
-- Anonymous

The fly that doesn't want to be swatted is most secure when it lights on the fly-swatter.
-- Georg C. Lichtenberg (1742-1799)

Forbidden things have a secret charm.
-- Pubilus Cornelius Tacitus (56 – 117 AD)

It is with trifles, and when he is off guard, that a man best reveals his character.
    -- Arthur Schopenhauer (1788-1860)

They are not all saints who use holy water.
    -- English proverb

Your religion is what you do when the sermon is over.
    -- Anonymous

All of us are experts at practicing virtue at a distance.
    -- Theodore M. Hesburgh

Level with your child by being honest. Nobody spots a phony quicker than a child.
    -- Mary MacCracken

Americans do not abide very quietly the evils of life.
    -- Richard Hofstadter (1916-1970)

In every American there is an air of incorrigible innocence, which seems to conceal a diabolical cunning.
    -- A. E. Housman (1859-1936)

Some people would be less dangerous if they had no good in them at all.
    -- François de La Rochefoucauld (1613-1680)

The only difference between the fool, and the criminal who attacks a system is that the fool attacks unpredictably and on a broader front.
    -- Tom Gilb

Call me paranoid but I don't trust spiders, I don't trust Predacons and I don't trust dames who sneak in and out of classified areas when they think that nobody is watching.
    -- From the movie, *Beast Wars: Transformers* (1996)

The only difference between saints and sinners is that every saint has a past while every sinner has a future.
    -- Oscar Wilde (1854-1900)

"I'd be glad to swear a loyalty oath. Hell, yes I'm loyal, you #*&~@$!"

Definition—Grawlix: The #*&~@$! symbols used in comics to represent swearing.

Definition—CAPTCHA: Visually distorted letters, numbers, or words that (ideally) only humans can read. Used on the Internet to establish that a human being is interacting with a web page, and not a software program. The term is an acronym for "Completely Asinine Public Turing test to tell Computers that Humans are Assholes".

Agents of disruption, subversion, sabotage and disinformation tunnelers and smugglers, listeners and forgers, trainers and recruiters and talent spotters and couriers and watchers and seducers, assassins and balloonists, lip readers and disguise artists.
   -- John LeCarre

Matt: Don't be cynical. Why do you always assume the worst about people?
Gwyn: Statistics.
   -- Dialog from the movie, *Miami Rhapsody* (1995)

Watch what people are cynical about, and one can often discover what they lack.
   -- George S. Patton (1885-1945)

What is the use of straining after an amiable view of things, when a cynical view is most likely to be the true one?
   -- George Bernard Shaw (1856-1950)

It is a sin to believe evil of others, but is seldom a mistake.
   -- H.L. Mencken (1880-1956)

It's amazing the clarity that comes with psychotic jealousy.
   -- From the movie, *My Best Friend's Wedding* (1997)

You can be a rank insider as well as a rank outsider.
   -- Robert Frost (1874-1963)

All money is tainted; tain't none of it mine.
   -- Thomas Francis McGuire

I can resist anything but temptation.
   -- Oscar Wilde (1854-1900)

There are several good protections against temptations, but the surest is cowardice.
   -- Mark Twain (1835-1910)

The most perfidious way of harming a cause consists of defending it deliberately with faulty arguments.
    -- Friedrich Nietzsche (1844-1900)

We have to distrust each other. It's our only defense against betrayal.
    -- Tennessee Williams (1911-1983)

Many a deep secret that cannot be pried out by curiosity can be drawn out by indifference.
    -- Sydney J. Harris (1917-1986)

If you would not step into the harlot's house, do not go by the harlot's door.
    -- Thomas Secker (1693-1768)

Knowledge is power, if you know it about the right person.
    -- Ethel Mumford (1878?-1940)

The chief lesson I have learned in a long life is that the only way to make a man trustworthy is to trust him; and the surest way to make him untrustworthy is to distrust him and show your distrust.
    -- Henry L. Stimson (1867-1950)

Amateurs hack systems, professionals hack people.
    -- Bruce Schneier

Things were run on a need-to-know basis; if you needed to know, you weren't told.
    -- Peter Jay on his boss at Maxwell Publishing

Self Respect: The secure feeling that no one, as yet, is suspicious.
    -- H.L. Mencken (1880-1956)

Motivations for insider attacks:
1. greed or severe financial need
2. revenge
3. terrorism
4. ideology, political activism, radicalism, or anarchism
5. coercion/blackmail
6. social engineering/seduction
7. narcissism or ego; the need to feel important, gain recognition, or be seen as clever
8. desire to prove that a warned about threat or vulnerability is real (Cassandra)

9. desire for excitement
10. mental illness(?)
11. inadvertent compromise of security through carelessness, human error, laziness, ignorance, disregard of good security practices, or arrogance

There are some who become spies for money, or out of vanity and megalomania, or out of ambition, or out of a desire for thrills. But the malady of our time is of those who become spies out of idealism.
-- Max Lerner (1902-1992)

As to the Seven Deadly Sins, I deplore Pride, Wrath, Lust, Envy, and Greed. Gluttony and Sloth I pretty much plan my day around.
-- Robert Brault

A mule will labor ten years willingly and patiently for you for the privilege of kicking you once.
-- William Faulkner (1897-1962)

There is no such a liar as an indignant man.
-- Friedrich Nietzsche (1844-1900)

What do you think spies are: priests, saints and martyrs? They're a squalid procession of vain fools, traitors too, yes; pansies, sadists and drunkards; people who play cowboys and Indians to brighten their rotten lives.
-- John le Carré

Hell has three gates: lust, anger, and greed.
-- The Bhagavad Gita

Heaven hath no rage like love to hatred turned, nor Hell a fury like a woman scorned.
-- William Congreve (1670-1729)

Chaperons don't enforce morality, they force immorality to be discreet.
-- Judith Martin

Asking for help is still one of the best social engineering tools for compromising security. There is an inherent desire for people to help other people.
-- Chris Hadnagy

8 - Human Resources & Security

Some Key Points:

• **The HR Department can be a powerful tool for security and for mitigating the insider threat.**

• **It most organizations, however, it makes security worse by serving as Secret Police, Judge, Jury, Torturer, and Executioner, and by not competently mitigating employee disgruntlement.**

• **The purpose of a grievance process is to reduce the insider threat and improve productivity—not to rubber stamp management blunders.**

Question on a job application form: Do you support the overthrow of the government by force, subversion, or violence? Answer from one applicant: Violence.

Few great men would have got past Personnel.
> -- Paul Goodman (1911–1972)

I don't hire anybody who's not brighter than I am. If they're not brighter than I am, I don't need them.
> -- Paul "Bear" Bryant (1913-1983)

When you go in for a job interview, ask if they ever press charges.
> -- Jack Handey

I am free of all prejudices. I hate everyone equally.
> -- Anonymous

Plata o Plomo: A phrase in Spanish which means "silver or lead", i.e., "accept a bribe or face assassination."

People don't leave jobs, they leave jerks.
> -- old adage

Girlfriend: Do you really wanna know?

Boyfriend: I asked, didn't I? I'm playing the role of concerned guy.
-- Dialog from the movie, *Mallrats* (1995)

The human-resources trade long ago proved itself, at best, a necessary evil—and at worst, a dark bureaucratic force that blindly enforces nonsensical rules, resists creativity, and impedes constructive change. HR is the corporate function with the greatest potential—the key driver, in theory, of business performance—and also the one that most consistently underdelivers.
-- Keith Hammonds. See http://www.fastcompany.com/magazine/97/open_hr.html?page=0,0

Not only does our Diversity Director not understand diversity, she wouldn't be in favor of it if she did!
-- Anonymous

You're too different to be on the diversity committee!
-- Actual accusation from a senior manager at Los Alamos National Laboratory trying to intimidate an employee into quitting the employee diversity committee

Anger is a signal, and one worth listening to.
-- Harriet Lerner

The purpose of the grievance process is to protect the institution.
-- HR employee

If a pig loses its voice, is it disgruntled?
-- Anonymous

The main purpose of a complaint resolution or grievance process should be to try to turn disgruntled employees into gruntled employees.
-- Roger Johnston

Regard your soldiers as your children, and they will follow you into the deepest valleys; look on them as your own beloved sons, and they will stand by you even in death.
-- Sun Tzu (544-496 BC)

Do not protect yourself by a fence, but rather by your friends.
-- Czech proverb

Employees will get more pissed off about not being consulted on the little things than they will on major new directions for the organization.
-- Anonymous

I consider myself to be a pretty good judge of people. That's why I don't like any of them.
-- Roseanne Barr

9 - The Adversary

Some Key Points:

• **Never underestimate your adversaries.**

• **You must get into their heads.**

• **The bad guys have most of the advantages because offense is easier than defense.**

• **Good security requires thinking about what the adversaries might do, and how to counter what they might do.**

When choosing between two evils, I always pick the one I never tried before.
-- Mae West (1893-1980)

I don't know of a greater advantage than to appreciate the worth of an enemy.
-- Johann Wolfgang von Goethe (1749-1832)

If you know the enemy and know yourself, you need not fear the results of a hundred battles.
-- Sun Tzu (544-496 BC)

So We're In Agreement Maxim: If you're happy with your security, so are the bad guys.

There is no little enemy.
-- Benjamin Franklin (1706-1790)

The bad guys are always going to be one step ahead of the good guys—they're more nimble, have less bureaucracy, are quicker to adapt to new technologies—and in a fast-changing technological world this gap is only going to get worse.
-- Bruce Schneier

Greed is for amateurs. Disorder, chaos, anarchy: now that's fun.
-- The Crow

There is always more spirit in attack than in defense.
-- Titus Livius (59 BC – 17 AD)

May the forces of evil become confused on the way to your house.
-- George Carlin (1937-2008)

Never wrestle with a pig. You can't win. You both get dirty. The pig loves it.
-- Attributed to Pasquale Capozzi

Friends come and go; enemies accumulate.
-- Anonymous

Your friends sometimes go to sleep; your enemies never do.
-- Thomas Brackett Reed (1839-1902)

And oftentimes, to win us to our harm,
The instruments of darkness tell us truths,
Win us with honest trifles, to betray's
In deepest consequence.
-- William Shakespeare (1564-1616), *Macbeth*, 1:3

No one will ever win the battle of the sexes; there's too much fraternizing with the enemy.
-- Henry Kissinger

The wise learn many things from their foes.
-- Aristophanes (446–386 BC)

You can discover what your enemy fears most by observing the means he uses to frighten you.
-- Eric Hoffer (1902-1983)

He Who's Name Must Never Be Spoken Maxim: Security programs and security professionals who don't talk a lot about "the adversary" or the "bad guys" aren't prepared for them and don't have good security. (From *Harry Potter*.)

My colleagues are spherical bastards. No matter how you look at them, they're bastards.
-- Cal Tech astronomer Fritz Zwicky (1989-1974)

The reverse side has a reverse side.

The Adversary

A truth that is told with bad intent beats all the lies you can invent.
   -- William Blake (1757-1827)

Hell is empty and all the devils are here.
   -- William Shakespeare (1564-1616), *The Tempest* 1.2

Hell is truth seen too late—duty neglected in its season.
   -- Tryon Edwards (1809-1894)

Hell isn't merely paved with good intentions, it is walled and roofed with them.
   -- Aldous Huxley (1894-1963)

Hell is a place where the motorists are French, the policemen are German, the cooks are English, the bureaucrats are Italian, and the lovers are Swiss.
   -- Anonymous

Morticia: I'm just like any modern woman trying to have it all. Loving husband, a family. It's just, I wish I had more time to seek out the dark forces and join their hellish crusade.
   -- From the movie, *Addams Family Values* (1993)

How does Bugs Bunny do it? How does he know when he wakes up in the morning to put in his pocket 3 sticks of dynamite, a physician costume, and a bicycle pump?
   -- Anonymous

The average man will bristle if you say his father was dishonest, but he will brag a little if he discovers that his great-grandfather was a pirate.
   -- Bern Williams

Why do hackers succeed? They're lucky, they're patient and they're brilliant. They're also better funded than you.
   -- John Stewart

She's as mean as a snake. She reminds me of me.
   -- Tennis player Martina Hingis

Nothing in the universe can travel at the speed of light, they say, forgetful of the shadow's speed.
   -- Howard Nemerov (1920-1991)

News Correction: In our cover story about Hunter S. Thompson yesterday, we mistakenly attributed to Richard Nixon the view that Hunter Thompson represented "that dark, venal and incurably violent side of the American character". On the contrary, it was Thompson who said that of Nixon.
> -- *The Guardian* (U.K.)

Yesterday upon the stair,
I met a man who wasn't there.
He wasn't there again today.
Oh how I wish he'd go away.
> -- Hughes Mearns (1875-1965)

10 - The Expected & the Unexpected

Some Key Points:

- **Expect the unexpected.**

- **We see what we expect to see, and miss what we are not prepared to see.**

- **You need the right mindset to have good security.**

If they expect us to expect the unexpected, doesn't the unexpected become the expected?
    -- Anonymous

To expect the unexpected shows a thoroughly modern intellect.
    -- Oscar Wilde (1854-1900)

Chance favors the prepared mind.
    -- Louis Pasteur (1822-1895)

We are never prepared for what we expect.
    -- James Michener (1907-1997)

Anything long expected take the form of the unexpected when at last it comes.
    -- Mark Twain (1835-1910)

He who is not prepared today will be less so tomorrow.
    -- Ovid (43 BC – 17 AD)

If you do not expect the unexpected, you will not find it; for it is hard to be sought out and difficult.
    -- Heraclitus (535 – 475 BC)

As a rule, we perceive what we expect to perceive. The unexpected is usually not perceived at all.
    -- Peter Drucker (1909-2005)

The eye sees only what the mind is prepared to comprehend.

-- Henri Bergson (1859-1941)

My sister's expecting a baby, and I don't know if I'm going to be an uncle or an aunt.
    -- Chuck Nevitt, North Carolina State basketball player, explaining to Coach
    Jim Valvano why he was so nervous during a game

11 - Makes Sense

Some Key Points:

• **The obvious is often more complex than you might imagine.**

Actual Headline: Man Accused of Pretending to Be a Lawyer Will Represent Himself.
-- Texarkana Gazette, March 8, 2011

I declare this thing open—whatever it is.
-- Prince Philip at the grand opening of an annex to the Vancouver City Hall

Welcome to President Bush, Mrs. Bush, and my fellow astronauts.
-- Dan Quayle at a ceremony for the 20th anniversary of the moon landing

Shouldn't the Air and Space Museum be empty?
-- Dennis Miller

I was still a shock when George died. It was the last thing I thought he'd do.
-- Angie Best

Never wear anything that panics the cat.
-- P.J. O'Rourke

Steve McQueen looks good in this movie. He must have made it before he died.
-- Yogi Berra (1925-2015)

"Product not actual size."
-- Disclaimer on a TV ad for Burger King showing a giant Whopper crushing a car

Wife: Before we were married, you said mother could stay with us whenever she pleased.
Husband: Yes, but she hasn't pleased yet.

Every minute was more exciting than the next.
-- Linda Evans

Never work with kids or animals.
-- W.C. Fields (1880-1946)

Who says nothing is impossible? I've been doing nothing for years.
-- Anonymous

Never judge a book by its movie.
-- J. W. Eagan

I'd like to say that you've been beaten by the best, but unfortunately it was only me.
-- Anonymous

What has posterity ever done for me?
-- Groucho Marx (1890-1977)

Of all the radio stations in Chicago…we're one of them.
-- Slogan of FM 105.9, a classic rock radio station in Chicago

Mel Brooks was once asked why so many TV comedy writers are Jewish. He responded that it's probably because their parents are.

Fathers send their sons to college either because they went to college or because they didn't.
-- L.L. Henderson

In the beginning there was nothing. God said, 'Let there be light!' And there was light. There was still nothing, but you could see it a whole lot better.
-- Ellen DeGeneres

When I told the people of Northern Ireland that I was an atheist, a woman in the audience stood up and said, "Yes, but is it the God of the Catholics or the God of the Protestants in whom you don't believe?"
-- Quentin Crisp (1909-1999)

Classified ad in a Northern Ireland newspaper:
WANTED: Man and woman to look after two cows, both Protestant.

On an American Airlines packet of nuts: "Instructions: Open packet, eat nuts."

On some Swann frozen dinners: "Serving suggestion: Defrost."

Makes Sense

She took an honorable icon that is seen in sporting venues everywhere and degraded it. Fortunately, the foam finger has been around long enough that it will survive this incident.
-- Steve Chmelar, inventor of the foam finger, after Miley Cyrus performed a lewd song and dance routine with one at the 2013 MTV Video Music Awards

Time flies like the wind. Fruit flies like a banana.
-- Groucho Marx (1890-1977)

If I had more time, I would write a shorter letter.
-- Blaise Pascal (1623-1662)

Be respectful to your superiors, if you have any.
-- Mark Twain (1835-1910), "Advice to Young People" speech, 4/15/1882

Hegel was right when he said that we learn from history that man can never learn anything from history.
-- George Bernard Shaw (1856-1950)

My parents didn't want to move to Florida, but they turned 60, and it was the law.
-- Jerry Seinfeld

Everything I did in my life that was worthwhile I caught hell for.
-- Supreme Court Justice Earl Warren (1891-1974)

If you are going to go through hell, come back having learned something.
-- Drew Barrymore

Printing in a church bulletin:
TONIGHT'S SERMON: WHAT IS HELL?
COME EARLY AND LISTEN TO OUR CHOIR PRACTICE.

They misunderestimated me.
-- George W. Bush

There's no such thing as fun for the entire family.
-- Jerry Seinfeld

We tend to scoff at the beliefs of the ancients. But we can't scoff at them personally, to their faces, and this is what annoys me.

-- Jack Handey

You know the world is going crazy when the best rapper is a white guy, the best golfer is a black guy, the tallest guy in the NBA is Chinese, the Swiss hold the America's Cup, France is accusing the U.S. of arrogance, Germany doesn't want to go to war, and the three most powerful men in America are named 'Bush', 'Dick', and 'Colon'.
-- Chris Rock

Heaven is a place where the cooks are French, the police are British, the mechanics are German, the lovers are Italian and everything is organized by the Swiss.
Hell is where the cooks are British, the police are German, the mechanics are French, the lovers are Swiss, and everything is organized by the Italians.
-- old joke

Reagan doesn't have that presidential look.
-- United Artists executive who rejected actor Ronald Reagan for the lead role in the 1964 film, *The Best Man*

The embarrassing thing is that the salad dressing is out-grossing my films.
-- Paul Newman (1925-2008)

The only man who had a proper understanding of Parliament was old Guy Fawkes.
-- George Bernard Shaw

Candy Corn is the only candy in the history of America that's never been advertised. And there's a reason. All of the candy corn that was ever made was made in 1911.
-- Louis Black

One day I was dangling a string for a cat to play with and I was thinking to myself, 'what a dumb cat. It's been chasing this string for an hour.' Then it hit me. I'd been dangling a string for an hour.
-- Anonymous

Remember, it doesn't matter whether you win or lose; what matters is whether I win or lose.
-- Darrin Weinberg

It's never just a game when you're winning.
-- George Carlin (1937-2008)

On the game show Family Fortunes:
Q: Name something easy to do forwards but difficult to do backwards.
Contestant: Eating.

Last night, I watched a porno so bad that I actually finished it.
   -- Anonymous

LIBRARY PARKING ONLY
Violators will be held in low esteem
   -- Local library parking sign

The difference between fiction and reality is that fiction has to make sense.
   -- Tom Clancy

I think it's wrong that only one company makes the game Monopoly.
   -- Steven Wright

The important thing in science is not so much to obtain new facts as to discover
new ways of thinking about them.
   -- William Bragg (1862-1942)

Science is a wonderful thing if one does not have to earn one's living at it.
   -- Albert Einstein (1879-1955)

Winston's Churchill's valet, Norman McGowan, was surprised one day to hear
Churchill muttering in his bath. McGowan thought that he was talking to him and
called out, "Do you want me, Sir?" "I wasn't talking to you, Norman," Winston
replied. "I was addressing the House of Commons."

I hate flowers—I paint them because they're cheaper than models and they don't
move.
   -- Georgia O'Keeffe: (1887-1986)

No one traveling on a business trip would be missed if he failed to arrive.
   -- Thorstein Veblen (1857-1929)

We've got a rule in the band—no matter what trouble you're going to get into, never
get arrested in a country that doesn't use your own alphabet. ... If you get arrested in
a country that uses squiggles, or a box or a line instead of proper letters, you're
fucked, mate, you're never coming home.

-- Noel Gallagher of *Oasis*

In what place and time has it ever been easy to be female?
-- Derek Worthington

Often, the [snack] chip aisle [at the grocery store] is disorganized and unappealing to women.
-- Gannon Jones

Actual legal testimony:
Q: You say it's Miss Jones. I take it then you are unmarried?
A: Yes, twice.

"Two mothers-in-law."
-- Lord John Russell (1832-1900), on being asked what he would consider proper punishment for bigamy.

If Barbie is so popular, why do you have to buy her friends?
-- Steven Wright

It's not Area 51 I'm worried about—it's Areas 1 through 50.
-- Anonymous

You've got to put square pegs in square holes—not the other way round.
-- Soccer player Steve Claridge

A special Providence protects fools, drunkards, small children, and the United States of America.
-- Otto von Bismarck (1815-1898)

America is the only country that went from barbarism to decadence without civilization in between.
-- Oscar Wilde (1854-1900)

Too bad the only people who know how to run the country are busy driving cabs and cutting hair.
-- George Burns (1896-1996)

Anyone who uses the phrase 'easy as taking candy from a baby' has never tried taking candy from a baby.
-- Anonymous

Channeling is just bad ventriloquism. You use another voice, but people can see your lips moving.
    -- Peter Jillette

Why do psychics have to ask you for your name?
    -- Steven Wright

Here's something to think about: How come you never see a headline like 'Psychic Wins Lottery'?
    -- Jay Leno

I almost had a psychic girlfriend but she left me before we met.
    -- Steven Wright

If you do things right, people won't be sure you've done anything at all.
    -- Anonymous

You've achieved success in your field when you don't know whether what you're doing is work or play.
    -- William Beatty

I'm not a member of any organized political party. I'm a Democrat!
    -- Will Rogers (1879-1935)

Someday, we'll look back on this, laugh nervously, and change the subject.
    -- Anonymous

When you travel, remember that a foreign country is not designed to make you comfortable. It is designed to make its own people comfortable.
    -- Clifton Fadiman (1904-1999)

You can't have a light without a dark to stick it in.
    -- Arlo Guthrie

The only reason I made a commercial for American Express was to pay for my American Express bill.
    -- Peter Ustinov (1921-2004)

Every single ad was based on some sort of dumb animal—be it a bear, pig, cow, dog, or supermodel.
    -- Kathleen Madigan on the 1987 Super Bowl commercials

If he raced his pregnant wife he'd finish third.
> -- Baseball manager Tommy Lasorda speaking about catcher Mike Scioscia

If you cannot get rid of the family skeleton, you may as well make it dance.
> -- George Bernard Shaw (1856-1950)

The gem cannot be polished without friction, nor man perfected without trials.
> -- Chinese proverb

Hood's Law: Never eat a cookie with a resistance of less than 1 ohm.

Actual Headline: New Dishwashers Leave Plates Clean Enough to Eat Off Of

Cremation Service: $895, plus a $35 one-time membership fee.
> -- Newspaper advertisement

The more you find out about the world, the more opportunities there are to laugh at it.
> -- Bill Nye

People are much too solemn about things—I'm all for sticking pins into episcopal behinds.
> -- Aldous Huxley (1894-1963)

It's a fine line between cleverness and stupidity.
> -- David St. Hubbins in the movie, *This is Spinal Tap* (1984)

When asked what he would do if he found a million dollars, Yogi Berra said, "I'd see if I could find the guy who lost it, and if he was poor, I'd give it back.

I don't like country music, but I don't mean to denigrate those who do. And for the people who like country music, 'denigrate' means to 'put down.
> -- Bob Newhart

Actual country music song titles:
I Gave Her the Ring and She Gave Me the Finger.
I Got in at 2 with a 10 and Woke up at 10 with a 2.
At the Gas Station of Love, I Got the Self-Service Pump.
Dog Poop On The Pillow Where Your Sweet Head Used To Be.
You Were Only A Splinter In My Ass As I Slid Down The Bannister Of Life.

Actual courtroom testimony:
Attorney: Do you know whether he put his seat belt on, or are you just surmising he didn't?
Witness: I know that he didn't put his seat belt on.
Attorney: What is your personal observation of that?
Witness: Because when we were driving down the street he was mooning people through the back window.

It's like deja vu all over again.
    -- Yogi Berra (1925-2015)

And there's fog on the M-25 in both directions.
    -- UK broadcaster John Humphrys

12 - Risk Management

Some Key Points:

• **Risk Assessment has to be creative and subjective.**

• **Risk Management is not about eliminating risk but (as the term implies) managing it, and being prepared to deal (via planning and resiliency) with security failures.**

• **The past is not a fully reliable guide to risk or the future.**

First weigh the considerations, then take the risks.
    -- Helmuth von Moltke (1800-1891)

The notion of a cold, analytic, actuarial risk assessment is largely a myth. Risk is a social construct that incorporates value judgments about context and cause.
    -- Henry Willis

Due to the rapid changes in the complexity of both technology and organizations over the past two decades, historical data has become less significant. Risk measurement and the identification of consequences require a combination of experience, skills, imagination, and creativity. This emphasis on subjective measurements is borne out in practice...
    -- David McNamee

Managing risk depends on accepting uncertainty; managing risk does not mean eliminating it.
    -- Robert Jamison (1774-1854)

In complex situations, we may rely too heavily on planning and forecasting and underestimate the importance of random factors in the environment. That reliance can also lead to delusions of control.
    -- Hillel J. Einhorn

Just Walk It Off Maxim: Most organizations will become so focused on prevention (which is very difficult at best) that they fail to adequately plan for mitigating attacks, and for recovering when attacks occur.

Let the fear of danger be a spur to prevent it; he that fears not, gives advantage to the danger.
  -- Muammar Qaddafi

If you don't risk anything, you risk even more.
  -- Erica Jong

The policy of being too cautious is the greatest risk of all.
  -- Jawaharial Nehru (1889-1964)

If no one ever took risks, Michelangelo would have painted the Sistine floor.
  -- Neil Simon

TV Host: Well, there's the first result, and the Silly Party has held Leicester. What do you make of that, Norman?
Norman: Well, this is largely as I predicted, except that the Silly Party won. I think this is largely due to the number of votes cast.
  -- Sketch from *Monty Python's Flying Circus*

The best security organizations I've worked with understand and attempt to quantify the risk-management decisions they make on an ongoing basis.
  -- Ed Moyle

You have to reach a level of comfort with that risk.
  -- Astronaut Sally Ride

There can be no vulnerability without risk; there can be no community without vulnerability; there can be no peace, and ultimately no life, without community.
  -- M. Scott Peck (1936-2005)

If we don't succeed, we run the risk of failure.
  -- Attributed to Dan Quayle

I'm afraid of sharks, but only in a water situation.
  -- Demetri Martin

The Associated Press reported that a college student in Oakwood, Georgia texted a friend that he would be near the West Hall high school. He meant to text, "Gunna be at West Hall this afternoon." But his smart phone's spell checker changed "Gunna" to "Gunman". Then he sent the text to the wrong phone number. The high school was locked down and general alarm ensured. (February 2012)

Life is not a journey to the grave with the intention of arriving safely in a pretty and well-preserved body. But rather, to skid in broadside, thoroughly used up, totally worn out, and loudly proclaiming .... WOW what a ride.
    -- Mark Frost

Let me tell you something you already know. The world ain't all sunshine and rainbows. It is a very mean and nasty place and it will beat you to your knees and keep you there permanently if you let it. You, me, or nobody is gonna hit as hard as life. But it ain't how hard you hit; it's about how hard you can get hit, and keep moving forward. How much you can take, and keep moving forward. That's how winning is done.
    -- Monolog from the movie, *Rocky Balboa* (2006)

13 - Vulnerability Assessments

Some Key Points:

• **Know thyself!**

• **Vulnerabilities are limitless. You can't find or eliminate them all.**

• **A new vulnerability assessment on the same security device, system, or program should find lots of new vulnerabilities. If not, go back and do it right with competent, independent assessors.**

• **Ask the right questions. Think like the bad guys.**

• **Finding flaws and attack scenarios is not enough. You must try to find possible fixes.**

• **Vulnerability fixes will introduce new vulnerabilities—hopefully less dangerous ones.**

• **Finding security vulnerabilities is good news (because you can then do something about them), not bad news!**

• **Perceived criticism is hard to take.**

• **Creativity is the key to a good vulnerability assessment.**

• **Watch out for "shoot the messenger"!**

• **Past security incidents cannot be your only guide to what might happen in the future.**

• **The bad guys get to define the problem, not the good guys!**

The greatest of faults, I should say, is to be conscious of none.
    -- Thomas Carlyle (1795-1881)

Man's greatest wisdom consists in knowing his own follies.

-- Madame de Sablé (1599-1678)

Once we know our weaknesses they cease to do us any harm.
    -- Georg C. Lichtenberg (1742-1799)

Our strength grows out of our weakness.
    -- Ralph Waldo Emerson (1803-1882)

It is a strength of character to acknowledge our failings and our strong points, and it is a weakness of character not to remain in harmony with both the good and the bad that is within us.
    -- Madame de Sablé (1599-1678)

He that wrestles with us strengthens our nerves and sharpens our skill. Our antagonist is our helper.
    -- Edmund Burke (1729-1797)

Mahbubani's Maxim: Organizations and security managers who cannot envision security failures will not be able to avoid them.

I cannot imagine any condition which could cause this ship to flounder. I cannot conceive of any vital disaster happening to this vessel.
    -- E.J. Smith, Captain of the Titanic, 1912

The purpose of a vulnerability assessment is to improve security. It is not a test you pass, a standard you meet, a welfare program for auditors, a rubber stamping of the status quo, justification for expenditures to date, or a technique for patting everybody on the back.
    -- Roger Johnston

A vulnerability assessment that finds no vulnerabilities is a waste of time and money. Go back and do it again with somebody who isn't incompetent, cowardly, or dishonest.
    -- Roger Johnston

There's always something new by looking at the same thing over and over.
    -- John Updike (1932-2009)

I was born to make mistakes, not to fake perfection.
    -- rapper Drake

A sudden, bold, and unexpected question doth many times surprise a man and lay him open.
-- Francis Bacon (1561-1626)

It is better to know some of the questions than all of the answers.
-- James Thurber (1894-1961)

Ask better questions. You'll get better answers.
-- Tom Monaham

Judge a man by his questions rather than his answers.
-- Voltaire (1694-1778)

There are no right answers to wrong questions.
-- Ursula K. Le Guin (1929-2018)

I've been watching a lot of game shows, and I've observed that the people with the answers come and go, but the man with the questions has a permanent job.
-- Gracie Allen (1895? – 1964)

An undefined problem has an infinite number of solutions.
-- Robert A. Humphrey

A problem well stated is a problem half solved.
-- Charles Franklin Kettering (1876-1958)

The hard part of doing a vulnerability assessment is not finding the security weaknesses. That is pretty easy if you are imaginative and think like the bad guys. The hard part is getting anybody to pay attention and to fix the problems.
-- Roger Johnston

If you're not getting criticized, you're not accomplishing anything.
-- Anonymous

Criticism is ... always a kind of complement.
-- John Maddox (1925-2009)

Honest criticism is hard to take, particularly from a relative, a friend, an acquaintance, or a stranger.
-- Franklin B. Jones

Before you criticize someone, walk a mile in his shoes. Then when you do criticize that person, you'll be a mile away and have his shoes.
    -- Billy Connolly

If you speak the truth, have a foot in the stirrup.
    -- Turkish proverb

Feynman's Maxim: An organization will fear and despise loyal vulnerability assessors and others who point out vulnerabilities or suggest security changes more than malicious adversaries.

A prophet is not honored in his own land.
    -- Traditional variation on Matthew 13:57

The prophet who fails to present a bearable alternative and yet preaches doom is part of the trap he postulates.
    -- Margaret Mead (1901-1978)

Don't find fault, find remedy.
    -- Henry Ford (1863-1947)

First Law of Revision: Information necessitating a change of design will be conveyed to the designers after—and only after—the plans are complete.

I don't have a solution, but I admire the problem.
    -- Anonymous

Peer's Law: The solution to the problem changes the problem.

This isn't right. This isn't even wrong.
    -- Wolfgang Pauli (1900-1958), commenting on a colleague's scientific paper

Your manuscript is both good and original, but unfortunately the part that is good is not original, and the part that is original is not good.
    -- Samuel Johnson (1709-1784)

I am returning this otherwise good typing paper to you because someone has printed gibberish all over it and put your name at the top.
    -- An English Professor at the University of Ohio commenting on his student's paper

The more original a discovery, the more obvious it seems afterwards.
    -- Arthur Koestler (1905-1983)

The Mona Lisa was stolen because nobody believed she could be.
    -- Francis Charmes (1848-1916)

Far more crucial than what we know or do not know is what we do not want to know. One often obtains a clue to a person's nature by discovering the reasons for his or her imperviousness to certain impressions.
    -- Eric Hoffer (1902-1983)

The history of Design Basis Threat in a nutshell: Take a common sense but nevertheless shocking idea (that security ought to be designed to counter the bad guys) and give it a gibberish sounding name to make it seem profound. Turn it into a big fad so that mind-numbingly unimaginative bureaucrats in many different organizations use it as a "tool" to justify the status quo. Next, allow the concept to be totally hijacked so that it becomes a bogus "test" of security, and only attacks that are arbitrarily assigned high probability merit consideration.
    -- Roger Johnston

He Who's Name Must Never Be Spoken Maxim: Security programs and professionals who don't talk a lot about "the adversary", "hackers", or the "bad guys" aren't prepared for them and don't have good security. (A reference to *Harry Potter*.)

We always prepare to fight the last war.
    -- Anonymous

It's a poor sort of memory that only works backwards.
    -- Lewis Carroll (1832-1898), *Alice in Wonderland*

There's nothing wrong with doing security surveys and audits; they're definitely useful. They just don't typically help you find any new vulnerabilities you haven't already envisioned. And bad guys don't do security surveys or audits, so these exercises won't help you think like the bad guys, see your security from their perspective, or predict how they might attack.
    -- Roger Johnston

It's perfectly reasonable to use the CARVER method or threat matrices. But they are not tools for finding vulnerabilities—as is often claimed. Instead, they are tools

for organizing one's thoughts about how and where to deploy security resources to deal with threats (and sometimes vulnerabilities). In other words, they are Risk Management tools, not Vulnerability Assessment tools *per se*.
-- Roger Johnston

Vulnerabilities Trump Threats Maxim: If you know the vulnerabilities (weaknesses), you've got a shot at understanding the threats (the probability that the weaknesses will be exploited, how, and by whom). Plus you might even be ok if you get the threats all wrong (which is possible). But if you focus only on the threats, you're probably in trouble.

Methodist Maxim: While vulnerabilities determine the methods of attack, most vulnerability or risk assessments will act as if the reverse were true.

Rigormortis Maxim: The greater the amount of rigor claimed or implied for a given security analysis, vulnerability assessment, risk management exercise, or security design, the less careful, clever, critical, imaginative, and realistic thought has gone into it.

I am Spartacus Maxim: Most vulnerability or risk assessments will let the good guys (and the existing security infrastructure, hardware, and strategies) define the problem, in contrast to real-world security applications where the bad guys get to.

I am not a pessimist; to perceive evil where it exists is, in my opinion, a form of optimism.
-- Roberto Rossellini (1906-1977)

When we were children, we used to think that when we were grown-up we would no longer be vulnerable. But to grow up is to accept vulnerability... To be alive is to be vulnerable.
-- Madeleine L'Engle (1918-2007)

Finding vulnerabilities is good news, not bad news. Vulnerabilities are always present in very large numbers. If we find one, it means we can do something about it. Finding a vulnerability doesn't typically mean somebody has been screwing up.
-- Roger Johnston

Is the glass half empty, half full, or twice as large as it needs to be?
-- Anonymous

Rat complaints have gone up, but we look at that as a positive thing, because more people know how to contact us now.
   -- New York City pest control bureaucrat

It's not a fair fight. The good guys have anal-retentive, bureaucratic idiots on their side who don't get rewarded even if they somehow manage (most likely through sheer dumb luck) to avoid a catastrophic security incident. The bad guys are smart, resourceful, and energetic. They get to have fun. And they get a big reward if they succeed.
   -- Anonymous

We need to find all the vulnerabilities.
   -- Delusional statement by a security bureaucrat

The number of different possible chess games is 10 raised to the 120th power. A player looking 8 moves ahead is already presented with as many possible games as there are stars in the galaxy. There are more possible chess games than the number of atoms in the universe.
   -- Garry Kasparov

It takes a thief to catch a thief.
   -- old English adage

If you want to be good, begin by assuming that you are bad.
   -- Epictetus (~50 to 120 AD)

It is sometimes expedient to forget who we are.
   -- Publilius Syrus (~42 BC)

I never criticise referees and I'm not going to change a habit for that prat.
   -- Ron Atkinson

I never questioned the integrity of an umpire. Their eyesight, yes.
   -- Leo Durocher (1905-1991)

I'm not allowed to comment on lousy officiating.
   -- Jim Finks, New Orleans Saints General Manager

Dare to be naive.
   -- Buckminster Fuller (1895-1983)

Everything you can imagine is real.
    -- Pablo Picasso (1881-1973)

Everything we see is inside our own heads.
    -- Buckminster Fuller (1895-1983)

Minds are like parachutes, they only function when open.
    -- Sir Thomas Dewar (1864-1930)

Thus, the task is not so much to see what no one has yet seen; but to think what nobody has yet thought, about that which everybody sees.
    -- Erwin Schrödinger (1887-1961)

Invader Zim: "You're lying! Nothing breaches my defenses, nothing! You hear me!....Nothing!" (To self: "Maybe there is some kind of flaw, but what?")
    -- Invader Zim TV cartoon, Episode 109b, *Rise of the Zitboy*

No Problem can be solved from the same consciousness that created it.
    -- Albert Einstein (1879-1955)

Reason cannot break out of its own loop.
    -- Mason Cooley (1927-2002)

It may well be doubted whether human ingenuity can construct an enigma of the kind which human ingenuity may not, by proper application, resolve.
    -- Edgar Allan Poe (1809-1849)

It is very unpleasant and annoying to see men, who claim to be peers of anyone in a certain field of study, take for granted certain conclusions which later are quickly and easily shown by another to be false.
    -- Salviati, in Galileo's *Two New Sciences* (1638)

If you look at life one way, there is always cause for alarm.
    -- Elizabeth Bowen (1899-1973)

An unexamined idea, to paraphrase Socrates, is not worth having and a society whose ideas are never explored for possible error may eventually find its foundations insecure.
    -- Mark Van Doren (1894-1972)

To find a fault is easy; to do better may be difficult.

-- Plutarch (46 AD – 120 AD)

If you are not criticized, you may not be doing much.
    -- Donald M. Rumsfeld

I'm just trying to make a smudge on the collective unconscious.
    -- David Letterman

Every wall is a door.
    -- Ralph Waldo Emerson (1803-1882)

We often get in quicker by the back door than by the front.
    -- Napoleon Bonaparte (1769-1821)

Anyone can see a forest fire. Skill lies in sniffing the first smoke.
    -- Robert Heinlein (1907-1988)

You cannot run away from weakness; you must some time fight it out or perish;
and if that be so, why not now, and where you stand?
    -- Robert Louis Stevenson (1850-1894)

To eat an egg, you must break the shell.
    -- Jamaican proverb

Men in authority will always think that criticism of their policies is dangerous. They
will always equate their policies with patriotism, and find criticism subversive.
    -- Henry Steele Commager (1902-1998)

To be able to think freely, a man must be certain that no consequence will follow
whatever he writes.
    -- Ernest Renan (1823-1892)

Where it is a duty to worship the sun it is pretty sure to be a crime to examine the
laws of heat.
    -- John Morley

Ideas are like rabbits. You get a couple and learn how to handle them, and pretty
soon you have a dozen.
    -- John Steinbeck (1902-1968)

The best way to have a good idea is to have lots of ideas.

-- Linus Pauling (1901-1994)

It had only one fault. It was kind of lousy.
    -- James Thurber (1894-1961)

It's easier to put on slippers than to carpet the whole world.
    -- Stuart Smalley

Sometimes security implementations look fool proof. And by that I mean proof that fools exist.
    -- Dan Philpott

Who am I to tamper with a masterpiece?
    -- Oscar Wilde (1854-1900), refusing to make alterations to one of his own plays

Talk sense to a fool and he calls you foolish.
    -- Euripides (~480 – ~406 BC)

I dip my pen in the blackest ink, because I'm not afraid of falling into my inkpot.
    -- Ralph Waldo Emerson (1803-1882)

Many well-meaning persons suppose that the discussion respecting the means for baffling the supposed safety of locks offers a premium for dishonesty, by showing others how to be dishonest. This is a fallacy. Rogues are very keen in their profession, and already know much more than we can teach them respecting their several kinds of roguery. Rogues knew a good deal about lockpicking long before locksmiths discussed it among themselves.
    -- Charles Tomlinson (1808-1987), *Rudimentary Treatise on the Construction of Locks*

It's not that the creative act and the critical act are simultaneous. It's more like you blurt something out and then analyze it.
    -- Robert Motherwell (1915 - 1991)

You unlock this door with the key of imagination. Beyond it is another dimension—a dimension of sound, a dimension of sight, a dimension of mind. You're moving into a land of both shadow and substance, of things and ideas. You've just crossed over into the Twilight Zone.
    -- Rod Serling (1924-1975), Introduction to the *Twilight Zone*

Low-Tech Maxim: Low-tech attacks work (even against high-tech devices and systems).

Do not touch anything unnecessarily. Beware of pretty girls in dance halls and parks who may be spies, as well as bicycles, revolvers, uniforms, arms, dead horses, and men lying on roads—they are not there accidentally.
    -- Soviet infantry manual from the 1930's

Criticism is often not a science; it is a craft, requiring more good health than wit, more hard work than talent, more habit than native genius. In the hands of a man who has read widely but lacks judgment, applied to certain subjects it can corrupt both its readers and the writer himself.
    -- Jean de la Bruyère (1645-1696)

To see what is in front of one's nose needs a constant struggle.
    -- George Orwell (1903-1950)

You can observe a lot by just watching.
    -- Yogi Berra (1925-2015)

A thing may look specious in theory, and yet be ruinous in practice; a thing may look evil in theory, and yet be in practice excellent.
    -- Edmund Burke (1729-1797)

It is sometimes useful to pretend we are deceived, because when we show a deceiving man that we see through his artifices, we only encourage him to increase his deceptions.
    -- Madame de Sablé (1599-1678)

Security is often about fixing blame. It should be focused on fixing problems.
    -- Anonymous

Whenever there is a simple error that most laymen fall for, there is always a slightly more sophisticated version of the same problem that experts fall for.
    -- Amos Tversky (1937-1996)

There is no error so monstrous that it fails to find defenders among the ablest men.
    -- John Dalberg-Acton (1834-1902)

How come wrong phone numbers are never busy?
    -- Anonymous

What was most significant about the lunar voyage was not that men set foot on the moon but that they set eye on the earth.
  -- Norman Cousins (1915-1990)

Umpire #1: I call 'em the way I see 'em.
Umpire #2: I call 'em the way they are.
Umpire #3: They ain't nothin' untils I calls 'em.

Out of nowhere the idea will appear. It will come to you when you least expect it.
  -- James Webb Young, *A Technique for Producing Ideas*

Could Hamlet have been written by committee, or the Mona Lisa painted by a club? Could the New Testament have been composed as a conference report? Creative ideas don't spring from groups. They spring from individuals.
  -- Alfred Whitney Griswold (1885-1959)

From now on, I'll connect the dots my own way.
  -- Bill Watterson, from *Calvin and Hobbes*

Some people see things that are and ask, Why?
Some people dream of things that never were and ask, Why not?
Some people have to go to work and don't have time for all that.
  -- George Carlin (1937-2008)

He that will not apply new remedies must expect new evils; for time is the greatest innovator.
  -- Francis Bacon (1561-1626)

The object in life is not to be on the side of the majority, but to be insane in such a useful way that they can't commit you.
  -- Mark Edwards

Only dead fish swim with the stream.
  -- Malcolm Muggeridge (1903-1990)

Factoid: A lake is not a depression where water accumulates. It's a spot where the ground drops below the water table.

Fantasies are more than substitutes for unpleasant reality; they are also dress rehearsals, plans. All acts performed in the world begin in the imagination.
  -- Barbara Grizzuti Harrison (1934-2002)

Great imaginations are apt to work from hints and suggestions and a single moment of emotion is sometimes sufficient to create a masterpiece.
-- Margaret Sackville (1881-1963)

The brain is wider than the sky.
For put them side by side,
The one the other will contain,
With ease, and you besides.
-- Emily Dickenson (1830-1886)

Creativity can solve almost any problem. The creative act, the defeat of habit by originality, overcomes everything.
-- George Lois

"I don't think that anybody could have predicted that these people would take an airplane and slam it into the World Trade Center, take another one and slam it into the Pentagon, that they would try to use an airplane as a missile ... even in retrospect there was nothing to suggest that.."
-- Testimony of Secretary of State Condoleezza Rice to the 9/11 Commission. The statement was later proven wrong, including by the 9/11 Commission in its report. Various foreign governments, intelligence analysts, scholars, terrorists, and FBI personnel had warned about this scenario prior to 9/11, and something similar was outlined in Tom Clancy's popular 1994 novel *Debt of Honor*.

Never solve a problem for someone, instead, help them figure out how to solve it on their own. Otherwise, you destroy their adaptive competence.
-- Lord Robin

No one wants advice—only corroboration.
-- John Steinbeck (1902-1968)

I always pass on good advice. It is the only thing to do with it. It is never of any use to oneself.
-- Oscar Wilde (1854-1900)

"High Security" is not the attribute of a product or security program. It's a context-dependent value judgment.
-- Roger Johnston

Definition—Vulnerability Assessors: Wiseguy troublemakers who appreciate the value of nothing, especially hard work.
-- *Devil's Dictionary of Security Terms*

14 - Creativity & Brainstorming

Some Key Points:

• Creativity is critical for good security and good vulnerability assessments.

• Individuals are creative, not groups. But groups are difficult to avoid when security is involved.

• Creativity depends on attitude and the environment.

• There are proven brainstorming techniques.

• A lot of what passes for "brainstorming" in organizations is just nonsense, not creative idea generation & problem solving.

• To be creative, lighten up!

• Individuals must be given ownership of their original idea and should be personally recognized for their creativity.

• The group environment needs to be diverse, high-energy, urgent (but not stressful), humorous/joyful/fun, cohesive but not too cohesive, competitive in a friendly and respectful manner, and enthusiastic about individual differences and eccentricities.

• Every idea, no matter how wacky or stupid, gets written down and treated as a gem, at least initially.

• Authority figures should not be involved, or at least should not act like authority figures.

• Most important: Keep the idea generation phase very separate from the criticizing, editing, or practical phase!

No one has ever had an idea in a dress suit.
    -- Frederick G. Banting (1891–1941)

It is impossible to see accurately how you look in your sunglasses.
    -- George Carlin (1937-2008)

Confidence in nonsense is a requirement for the creative process.
    -- Anonymous

There is a correlation between the creative and the screwball. So we must suffer the screwball gladly.
    -- Kingston Brewster

Happy people tend to be more creative problem solvers.
    -- Joseph T. Hallinan

Sanity is a one trick pony—all you have is rational thought. But when you're good and loony, the sky's the limit!
    -- The Tick

You're only given a little spark of madness. You mustn't lose it.
    -- Robin Williams

I never came upon any of my discoveries through the process of rational thinking.
    -- Albert Einstein (1879-1955)

The compulsion to take ourselves seriously is in inverse proportion to our creative capacity. When the creative flow dries up, all we have left is our importance.
    -- Eric Hoffer (1902-1983)

Creativity comes from trust. Trust your instincts.
    -- Rita Mae Brown

The worst enemy to creativity is self-doubt.
    -- Sylvia Plath (1932-1963)

The essential part of creativity is not being afraid to fail.
    -- Edwin H. Land (1909-1991)

Creativity is the ability to see relationships where none exist.
    -- Thomas Disch

Creativeness often consists of merely turning up what is already there. Did you know that right and left shoes were thought up only a little more than a century ago?

-- Bernice Fitz-Gibbon (1894-1992)

Nothing can inhibit and stifle the creative process more—and on this there is unanimous agreement among all creative individuals and investigators of creativity—than critical judgment applied to the emerging idea at the beginning stages of the creative process. ... More ideas have been prematurely rejected by a stringent evaluative attitude than would be warranted by any inherent weakness or absurdity in them. The longer one can linger with the idea with judgment held in abeyance, the better the chances all its details and ramifications [can emerge].
   -- Eugene Raudsepp (1923-1995)

Peer review and specialization are the worst things for creativity. They completely militate against working outside very narrow parameters.
   -- Mark C. Taylor

"Out of the box" problem-solvers have developed a series of habits to connect the dots effortlessly and trigger creativity frequently in order to solve problems elegantly.
   -- Pearl Zhu

The good ideas are all hammered out in agony by individuals, not spewed out by groups.
   -- Charles Brower (1857-1924)

There is evidence that brainstorming sometimes works better with nominal groups—pooled results from individual brainstormers—than with actual ones.
   -- Raymond S. Nickerson

The creative individual...is capable of more wisdom and virtue than collective man ever can be.
   -- John Stuart Mill (1806-1873)

Our species is the only creative species, and it has only one creative instrument, the individual mind and spirit of a man. Nothing was ever created by two men. There are no good collaborations, whether in music, in art, in poetry, in mathematics, in philosophy. Once the miracle of creation has taken place, the group can build and extend it, but the group never invents anything. The preciousness lies in the lonely mind of a man.
   -- John Steinbeck (1902-1968)

If you happen to be one of the fretful minority who can do creative work, never force an idea; you'll abort it if you do. Be patient and you'll give birth to it when the time is ripe. Learn to wait.
    -- Robert Heinlein (1907-1988)

A new idea is delicate. It can be killed by a sneer or a yawn; it can be stabbed to death by a joke or worried to death by a frown on the right person's brow.
    -- Charles Brower (1857-1924)

Creativity is so delicate a flower that praise tends to make it bloom, while discouragement often nips it in the bud. Any of us put out more and better ideas if our efforts are truly appreciated.
    -- Alexander Osborn (1888-1966)

Sometimes you've got to let everything go—purge yourself. If you are unhappy with anything....whatever is bringing you down, get rid of it. Because you'll find that when you're free, your true creativity, your true self comes out.
    -- Singer Tina Turner

Very few people do anything creative after the age of thirty-five. The reason is that very few people do anything creative before the age of thirty-five.
    -- Joel Hildebrand (1881-1983)

Creativity is inventing, experimenting, growing, taking risks, breaking rules, making mistakes, and having fun.
    -- Mary Lou Cook

Don't think. Thinking is the enemy of creativity. It's self-conscious, and anything self-conscious is lousy. You can't try to do things. You simply must do things.
    -- Ray Bradbury (1920-2012)

We all know your idea is crazy. The question is, is it crazy enough?
    -- Niels Bohr (1885-1962)

It is generally recognized that creativity requires leisure, an absence of rush, time for the mind and imagination to float and wander and roam, time for the individual to descend into the depths of his or her psyche, to be available to barely audible signals rustling for attention. Long periods of time may pass in which nothing seems to be happening. But we know that kind of space must be created if the mind is to leap

out of its accustomed ruts, to part from the mechanical, the known, the familiar, the standard, and generate a leap into the new.
    -- Nathaniel Branden

Any activity becomes creative when the doer cares about doing it right, or doing it better.
    -- John Updike (1932-2009)

It's really cool to just be really creative and create something really cool.
    -- Britney Spears

To stimulate creativity, one must develop the childlike inclination for play and the childlike desire for recognition.
    -- Albert Einstein (1879-1955)

A hunch is creativity trying to tell you something.
    -- Anonymous

When management schedules a "brainstorming" session, it's not true brainstorming.
    -- Anonymous

Imagination enables us to create a more exalting and consoling nature than what just a glance at reality allows us to perceive.
    -- Vincent van Gogh (1853-1890)

Good taste is the enemy of creativity.
    -- Pablo Picasso (1881-1973)

Extroverts are more likely to take a quick-and-dirty approach to problem-solving, trading accuracy for speed, making increasing numbers of mistakes as they go, and abandoning ship altogether when the problem seems too difficult or frustrating. Introverts think before they act, digest information thoroughly, stay on task longer, give up less easily, and work more accurately. Introverts and extroverts also direct their attention differently: if you leave them to their own devices, the introverts tend to sit around wondering about things, imagining things, recalling events from their past, and making plans for the future. The extroverts are more likely to focus on what's happening around them. It's as if extroverts are seeing "what is" while their introverted peers are asking "what if."
    -- Susan Cain

The higher you soar, the smaller you look to those who cannot fly.
    -- Anonymous

In this country, we encourage "creativity" among the mediocre, but real bursting creativity appalls us. We put it down as undisciplined, as somehow "too much".
    -- Pauline Kael (1919-2001)

Calvin: You can't just turn on creativity like a faucet. You have to be in the right mood.
Hobbes: What mood is that?
Calvin: Last-minute panic.
    -- Bill Watterson, from *Calvin and Hobbes*

Be brave enough to live creatively. The creative is the place where no one else has ever been. You have to leave the city of your comfort and go into the wilderness of your intuition. You cannot get there by bus, only by hard work, risking, and by not quite knowing what you are doing.
    -- Alan Alda

15 - Cognitive Dissonance & the Lack of Intellectual Humility—the greatest threat

Some Key Points:

• **Enemy #1 is you.**

• **"Cognitive dissonance" is the mental feeling of tension when there is a mismatch between what we want to be true and what is likely to be true. A security manager may desperately want her security to be effective, but she may be presented with evidence that there are serious problems. Unless cognitive dissonance is properly controlled, it can lead to denial, wishful thinking, and self-justification (self-serving rationalization and excuse making), paralysis or stagnation (failure to confront serious problems or take necessary actions), and confirmation bias or motivated reasoning (unduly dismissing ideas, arguments, evidence, or data that might call into question our current viewpoints, strong hopes, or past decisions). Poor security is often found in organizations with a lot of cognitive dissonance, and especially those that lack the security culture or mental discipline to handle it.**

• **Good security is not compatible with over-confidence or arrogance.**

• **If you're not scared about your security, you are making a mistake.**

• **Watch out for groupthink, blind allegiance to the majority view, conceit, wishful thinking, denial, the absence of doubt, dogmatic or binary thinking, and "shooting the messenger"!**

The truth? You can't handle the truth.
   -- From Jack Nicholson's speech in the movie, *A Few Good Men* (1992)

As scarce as the truth is, the supply has always been in excess of the demand.
   -- Josh Billings (1818-1885)

Speak Truth to Power.
   -- Traditional Quaker adage

It is impossible for a man to learn what he thinks he already knows.
-- Epictetus (~55 - ~135 AD)

The truth that is suppressed by friends is the readiest weapon of the enemy.
-- Robert Louis Stevenson (1850-1894)

Man will occasionally stumble over the truth, but most of the time he will pick himself up and continue on.
-- Winston Churchill (1874-1965)

Believe those who are seeking truth, doubt those who find it.
-- Andre Gide

The truth will set you free, but first it will piss you off.
-- Mal Pancoast

Truth does less good in the world than the appearance of truth does evil.
-- François de La Rochefoucauld (1613-1680)

Whenever, therefore, people are deceived and form opinions wide of the truth, it is clear that the error has slid into their minds through the medium of certain resemblances to that truth.
-- Socrates (469 – 399 BC)

It is the mark of an educated mind to be able to entertain a thought without accepting it.
-- Aristotle (384 - 322 BC)

I happen to feel that the degree of a person's intelligence is directly reflected by the number of conflicting attitudes she can bring to bear on the same topic.
-- Lisa Alther

The test of a first-rate intelligence is the ability to hold two opposed ideas in the mind at the same time, and still retain the ability to function.
-- F. Scott Fitzgerald (1896–1940)

I have opinions of my own—strong opinions—but I don't always agree with them.
-- George H.W. Bush

Sometimes when I close my eyes, I can't see.
-- Anonymous

Doubt is not a very pleasant condition, but certainty is absurd.
   -- Voltaire (1694-1778)

He is the wisest philosopher who holds his theory with some doubt.
   -- Michael Faraday (1791-1867)

Only fools are positive.
   -- Moe Howard (1897-1975), actor with the Three Stooges

To be positive: To be mistaken at the top of one's voice.
   -- Ambrose Bierce (1842–1914?), *The Devil's Dictionary*

You risk just as much in being credulous as in being suspicious.
   -- Denis Diderot (1713-1784)

Denial ain't just a river in Egypt.
   -- Mark Twain (1835-1910)

Never believe anything until it has been officially denied.
   -- Claud Cockburn (1904-1981)

It's not denial. I'm just selective about the reality I accept.
   -- Bill Watterson, from *Calvin and Hobbes*

The first step in the risk management process is to acknowledge the reality of risk. Denial is a common tactic that substitutes deliberate ignorance for thoughtful planning.
   -- Charles Tremper

Delay is the deadliest form of denial.
   -- C. Northcote Parkinson (1906-1993)

There is no stigma attached to recognizing a bad decision in time to install a better one.
   -- Laurence J. Peter (1919-1988)

Galileo's Maxim: The more important the assets being guarded, or the more vulnerable the security program, the less willing its security managers will be to hear about vulnerabilities. (The name of this maxim comes from the 1633 Inquisition where Church officials refused to look into Galileo's telescope out of fear of what they might see.)

"We want to follow a middle path between due diligence and paranoia."
-- Which really means, "We don't want to hear about any damn vulnerabilities!"

There are some things so serious you have to laugh at them.
-- Niels Bohr (1885-1962)

I don't want any yes-men around me. I want everyone to tell me the truth—even if it costs him his job.
-- Samuel Goldwyn (1879-1974)

The President doesn't want any yes-men and yes-women around him. When he says no, we all say no.
-- Elizabeth Dole, aide to President Reagan

If a man is offered a fact which goes against his instincts, he will scrutinize it closely, and unless the evidence is overwhelming, he will refuse to believe it. If, on the other hand, he is offered something which affords a reason for acting in accordance to his instincts, he will accept it even on the slightest evidence.
-- Bertrand Russell (1872-1970)

People want to believe, and as soon as they want to believe, they stop thinking.
-- Scott Wolter

When my information changes, I alter my conclusions. What do you do, sir?
-- attributed to John Maynard Keynes (1883-1946)

Self knowledge is always bad news.
-- John Barth

Americans are suckers for good news.
-- Adlai Stevenson (1900-1965)

The pendulum of the mind oscillates between sense and nonsense, not between right and wrong.
-- Carl Jung (1875-1961)

If you don't disagree with me, how will I know I'm right?
-- Samuel Goldwyn (1879-1974)

Refusal to believe until proof is given is a rational position; denial of all outside of our own limited experience is absurd.
    -- Annie Besant (1847-1933)

With most people disbelief in a thing is founded on a blind belief in some other thing.
    -- Georg C. Lichtenberg (1742-1799)

Few people are wise enough to prefer useful criticism over treacherous flattery.
    -- François de La Rochefoucauld (1613-1680)

We accept the reality of the world with which we are presented.
    -- From the movie, *The Truman Show* (1998)

What's offensive is that I'm portrayed as this prima donna with these sycophants telling me how great I am all the time. Yes, they do work for me, but we're working together for the greater good.
    -- Demi Moore

The smaller the mind the greater the conceit.
    -- Aesop (620 – 560 BC)

I have great faith in fools; self-confidence my friends call it.
    -- Edgar Allan Poe (1809-1849)

Stop the habit of wishful thinking and start the habit of thoughtful wishes.
    -- Mary Martin (1913-1990)

Always acknowledge a fault. This will throw those in authority off their guard and give you an opportunity to commit more.
    -- Mark Twain (1835-1910)

The aim of flattery is to soothe and encourage us by assuring us of the truth of an opinion we have already formed about ourselves.
    -- Edith Sitwell (1887-1964)

People ask you for criticism, but they only want praise.
    -- W. Somerset Maugham (1874-1965)

Confidence is that feeling you sometimes have before you fully understand the situation.

-- Anonymous

Danger breeds best on too much confidence.
    -- Pierre Corneille (1606-1684)

Confidence cannot find a place wherein to rest in safety.
    -- David Tuvill

Overconfidence is greatest in our own area of expertise.
    -- Edward Dolnick

Overconfidence is a leading cause of human error.
    -- Joseph T. Hallinan

Confidence is always over confidence.
    -- Robert Byrne

If you get people to play devil's advocate with themselves—asking what the evidence against this is—overconfidence is pretty close to being eliminated.
    -- Paul Shoemaker

An apology for the Devil: It must be remembered that we have only heard one side of the case. God has written all the books.
    -- Samuel Butler (1612-1680)

It is the assumptions that you believe the most deeply or that have held as true for the longest time that are likely to prove your undoing.
    -- Matthew S. Olson and Derek Van Bever

The fact that an opinion has been widely held is no evidence whatever that it is not utterly absurd.
    -- Bertrand Russell (1872-1970)

Whenever you find yourself on the side of the majority, it is time to pause and reflect.
    -- Mark Twain (1835-1910)

Everything popular is wrong.
    -- Oscar Wilde (1854-1900)

It would be silly to argue that the conventional wisdom is never true. But noticing where the conventional wisdom may be false—noticing, perhaps the contrails of sloppy or self-interested thinking—is a nice place to start asking questions.
-- Steven D. Levitt & Stephen J. Dubner

Factoid: The term "conventional wisdom" was coined by economist John Kenneth Galbraith (1908-2006).

Symptoms of Groupthink: closed-mindedness, pressure towards uniformity, overestimation of the group's or organization's abilities, limited consideration of alternatives and dissonant information.
-- Matthew S. Olson and Derek Van Bever

In the multitude of counselors there is safety.
-- Proverbs 11:14

Diversity of thought, not just of demographics is a good thing.
-- Anonymous

There is no housing shortage in Lincoln today—just a rumor that is put about by people who have nowhere to live.
-- G.L. Murfin, Mayor of Lincoln

New ideas stir from every corner. They show up disguised innocently as interruptions, contradictions and embarrassing dilemmas. Beware of total strangers and friends alike who shower you with comfortable sameness, and remain open to those who make you uneasy, for they are the true messengers of the future.
-- Rob Lebow

The most important scientific revolutions all include, as their only common feature, the dethronement of human arrogance from one pedestal after another of previous convictions about our centrality in the cosmos.
-- Stephen Jay Gould (1941-2002)

We associate truth with convenience, with what most closely accords with self-interest and personal well-being or promises best to avoid awkward effort or unwelcome dislocation of life. We also find highly acceptable what contributes most to self-esteem.
-- John Kenneth Galbraith (1908-2006)

Rig the Rig Maxim: Any supposedly "realistic" test of security is rigged.

Any man worth his salt will stick up for what he believes right, but it takes a slightly bigger man to acknowledge instantly and without reservation that he is in error.
-- Peyton C. March (1864-1955)

When people are engaged in something they are not proud of, they do not welcome witnesses. In fact, they come to believe the witness causes the trouble.
-- John Steinbeck (1902-1968)

Nothing is more intolerable than to have to admit to yourself your own errors.
-- Ludwig Van Beethoven (1770-1827)

It is difficult to get a man to understand something when his salary depends on his not understanding it.
-- Upton Sinclair (1878-1968)

Nothing is easier than self-deceit. For what each man wishes, that he also believes to be true.
-- Demosthenes (384 – 322 BC)

Doubt, indulged and cherished, is in danger of becoming denial; but if honest, and bent on thorough investigation, it may soon lead to full establishment of the truth.
-- Ambrose Bierce (1842-1914?)

Be not ashamed of mistakes and thus make them crimes.
-- Confucius (551 – 479 BC)

If you look for truth, you may find comfort in the end; if you look for comfort you will get neither truth nor comfort...only soft soap and wishful thinking to begin, and in the end, despair.
-- C.S. Lewis (1898-1963)

If a man will begin with certainties, he shall end in doubts; but if he will be content to begin with doubts he shall end in certainties.
-- Francis Bacon (1561-1626)

Men do not like to admit to even momentary imperfection. My husband forgot the code to turn off the alarm. When the police came, he wouldn't admit he'd forgotten the code...he turned himself in.
-- Rita Rudner

To most of us nothing is so invisible as an unpleasant truth. Though it is held before our eyes, pushed under our noses, rammed down our throats—we know it not.
    -- Eric Hoffer (1902-1983)

The ultimate security is your understanding of reality.
    -- H. Stanley Judd

Skepticism, like chastity, should not be relinquished too readily.
    -- George Santayana (1863-1952)

Failure occurs when you do not conceive of failure.
    -- Kishore Mahbubani

Fear is a reaction you have when you are getting closer to the truth.
    -- Jim Palmer

There are all kinds of devices invented for the protection and preservation of countries: defensive barriers, forts, trenches, and the like... But prudent minds have as a natural gift one safeguard which is the common possession of all, and this applies especially to the dealings of democracies. What is this safeguard? Skepticism. This you must preserve. This you must retain. If you can keep this, you need fear no harm.
    -- Demosthenes (384 – 322 BC)

To have doubted one's own first principles is the mark of a civilized man.
    -- Oliver Wendell Holmes, Jr. (1841–1935)

The most common of all follies is to believe passionately in the palpably not true. It is the chief occupation of mankind.
    -- H.L. Mencken (1880-1956)

To me a real patriot is like a real friend. Who's your real friend? It's the person who tells you the truth. That's who my real friends are. So, you know, I think as far as our country goes, we need more people who will do that.
    -- Bill Maher

I do think the patriotic thing to do is to critique my country. How else do you make a country better but by pointing out its flaws?
    -- Bill Maher

True friends stab you in the front.
    -- Oscar Wilde (1854-1900)

We are never deceived; we deceive ourselves.
    -- Johann Wolfgang von Goethe (1749-1832)

If you're afraid to let someone else see your weakness, take heart: Nobody's perfect. Besides, your attempts to hide your flaws don't work as well as you think they do.
    -- Julie Morgenstern

Take the attitude of a student, never be too big to ask questions, never know too much to learn something new.
    -- Og Mandino (1923–1996)

If you would be a real seeker after truth, it is necessary that at least once in your life you doubt, as far as possible, all things.
    -- Rene Descartes (1596–1650)

The greatest conflicts are not between two people but between one person and himself.
    -- Country music star Garth Brooks

The problem is not that there are problems. The problem is expecting otherwise and thinking that having problems is a problem.
    -- Theodore Rubin

The problem is not the problem, the problem is your attitude about the problem.
    -- Captain Jack Sparrow from the movie, *Pirates of the Caribbean*

If I only had a little humility, I'd be perfect.
    -- Ted Turner

You can fool too many of the people too much of the time.
    -- James Thurber (1894-1961)

This man is walking down the street when he spots another guy, stops, and speaks to him. "Whippledorn!" he says, "I haven't seen you in ages. It's amazing, you've changed so much! You've lost 30 pounds, gained 4 inches in height, and changed your hair color." The other man looks confused and says, "Look, my name is not Whippledorn!" "Wow," says the first man, "you've even changed your name!"

Mr. Spock Maxim: The effectiveness of a security device, system, or program is inversely proportional to how angry or upset people get about the idea that there might be vulnerabilities.

Where we have strong emotions, we're liable to fool ourselves.
-- Carl Sagan (1934-1996)

I don't want to believe. I want to know.
-- Carl Sagan (1934-1996)

When you are certain you cannot be fooled, you become easy to fool.
-- Edward Teller (1908-2003)

The first principle is that you must not fool yourself—and you are the easiest person to fool.
— Richard Feynman (1918-1988)

Scientists are the easiest to fool. They think in straight, predictable, directable, and therefore misdirectable, lines. The only world they know is the one where everything has a logical explanation and things are what they appear to be. Children and conjurors—they terrify me. Scientists are no problem; against them I feel quite confident.
-- Spoken by Zambendorf in *Code of the Lifemaker*, (James Hogan, 1987)

Lying to ourselves is more deeply ingrained than lying to others.
-- Fyodor Dostoyevsky (1821-1881)

It ain't what you don't know that gets you into trouble. It's what you know for sure that just ain't so.
-- Mark Twain (1835-1910)

The usual risks: what you don't know, what you're not thinking about, and (probably the biggest risk) what you are 100% sure about [that isn't true].
-- Bruce Berkowitz

To show resentment at a reproach is to acknowledge that one may have deserved it.
-- Tacitus (55-117 AD)

Nothing more completely baffles one who is full of trick and duplicity than straightforward and simple integrity in another. A knave would rather quarrel with a brother knave than with a fool, but he would rather avoid a quarrel with one honest man than with both. He can combat a fool by management and address, and he can conquer a knave by temptations. But the honest man is neither to be bamboozled nor bribed.
  -- C.C. Colton (1780-1832)

Mermaid Maxim: The most common excuse for not fixing security vulnerabilities is that they simply can't exist.

Onion Maxim: The second most common excuse for not fixing security vulnerabilities is that "we have many layers of security", i.e., we rely on "Security in Depth".

Hopeless Maxim: The third most common excuse for not fixing security vulnerabilities is that "all security devices, systems, and programs can be defeated". (Often invoked by the same person who previously believed the Mermaid Maxim after a vulnerability is directly demonstrated to him.)

Takes One to Know One Maxim: The fourth most common excuse for not fixing security vulnerabilities is that "our adversaries are too stupid and/or unresourceful to figure that out."

Thursday Maxim: Organizations and security managers will tend to automatically invoke irrational or fanciful reasons for claiming that they are immune to any postulated or demonstrated attack. (The name of this maxim refers to the idea that if a vulnerability or attack is demonstrated on Tuesday, it will be thought of as not applying on Thursday.)

Prejudices are what fools use for reason.
  -- Voltaire (1694-1778)

The infinitely little have a pride infinitely great.
  -- Voltaire (1694-1778)

There is only one pretty child in the world, and every mother has it.
  -- Chinese proverb

"No, no!" said the Queen. "Sentence first—verdict afterwards."
  -- *Alice's Adventures in Wonderland*

A few months ago, I told the American people I did not trade arms for hostages. My heart and my best intentions still tell me that's true, but the facts and the evidence tell me it is not.
-- President Ronald Reagan (1911-2004)

Close your eyes and wonder why it's dark.
Built a house of cards around your heart.
Act surprised when you keep fallin' down.
Something about you is like a spark.
Didn't think you'd tear it all apart
-- Joan Jett Lyrics from the song *Naked*

A man should never be ashamed to own he has been in the wrong, which is but saying... that he is wiser today than he was yesterday.
-- Alexander Pope (1688-1744)

It's only when you look at an ant through a magnifying glass on a sunny day that you realize how often they burst into flames.
-- Harry Hill

When you make a mistake, admit it. If you don't, you only make matters worse.
-- Ward Cleaver

With knowledge comes more doubt.
-- Johann Wolfgang von Goethe

Shrink: I'd like to show you some ink blots and have you tell me what you see.
Patient: OK.
Shrink: Here is ink blot #1.
Patient: A man and a woman making love in the back of a van.
Shrink: Here is ink blot #2.
Patient: Two naked girls making out in the shower.
Shrink: #3.
Patient: A threesome really going at it.
Shrink: Well, I think I understand your problem. You're obsessed with sex!
Patient: Me? You're the one with all the dirty pictures!

16 - Security by Obscurity

Some Key Points

- **Security based on keeping long-term secrets does not work.**

- **Somewhat counter-intuitively, security is usually better when it is reasonably transparent.**

Two can keep a secret if one is dead.
    -- Benjamin Franklin (1706-1790)

Three may keep a secret, if two of them are dead.
    -- Benjamin Franklin (1706-1790)

To keep your secret is wisdom; but to expect others to keep it is folly.
    -- Samuel Johnson (1709-1784)

Shannon's (Kerckhoffs') Maxim: The adversaries know and understand the security hardware and strategies being employed.

Corollary to Shannon's Maxim: Thus, "Security by Obscurity", i.e., security based on keeping long-term secrets, is not a good idea.

As a result of the Valerie Plume scandal in 2003 in which Administration sources leaked the name of a CIA operative to the press, President George W. Bush issued secret orders to end all Administration leaks of sensitive information. That order was promptly leaked to the press.

The vanity of being known to be trusted with a secret is generally one of the chief motives to disclose it.
    -- Samuel Johnson (1709-1784)

Everything secret degenerates … nothing is safe that does not show how it can bear discussion and publicity.
    -- Attributed to Lord Acton (1834-1902)

Nothing is so burdensome as a secret.

-- French proverb

Anything will give up its secrets if you love it enough. Not only have I found that when I talk to the little flower or to the little peanut they will give up their secrets, but I have found that when I silently commune with people they give up their secrets also—if you love them enough.
    -- George Washington Carver (1864-1943)

Secrecy has many advantages, for when you tell someone the purpose of any object right away, they often think there is nothing to it.
    -- Johann Wolfgang von Goethe (1749-1832)

If you wish to preserve your secret, wrap it up in frankness.
    -- Alexander Smith

The man who can keep a secret may be wise, but he is not half as wise as the man with no secrets to keep.
    -- Edgar Watson Howe (1853-1937)

A secret is your slave when you keep it, your master if you lose it.
    -- Arabian proverb

To know that one has a secret is to know half the secret itself.
    -- Henry Ward Beecher (1813-1887)

If you let the cat out of the bag never try to cram it back in again. This only makes matters worse.
    -- Anonymous

The big lie of computer security is that security improves by imposing complex passwords on users. In real life, people write down anything they can't remember. Security is increased by designing for the way humans actually behave.
    -- Jakob Nielsen

Definition—secret password: Any password that you forget because nobody can remember 16 gibberish characters.

It's ironic and counter-intuitive, but security is usually best when it is fairly transparent. This allows for review, criticism, analysis, improvement, accountability, and employee buy-in. Besides, the secrets you think you are keeping are probably not all that secret, especially if they are long-term secrets.

Security Sound Bites

-- Roger Johnston

17 - Security in Depth/Layered Security

Some Key Points:

• **"Defense in Depth" = "Security in Depth" = "Layered Security"**

• **Both a useful tool AND frequently a major impediment to good security (second only to cognitive dissonance).**

Definition—Security in Depth: We're desperately hoping that multiple layers of lousy security will somehow magically add up to good security.

Depth, What Depth? Maxim: For any given security program, the amount of critical, imaginative, skeptical, and intelligent thinking that has been undertaken is inversely proportional to how strongly the strategy of "Security in Depth" (layered security) is embraced.

We don't need to worry about this vulnerability because we have "security in depth".
     -- Traditional delusional cop-out of security bureaucrats everywhere

How many sieves do you have to stack before the water won't leak through?
     -- Anonymous

Factoid: The Maginot Line in World War II was an example of defense in depth. It was totally ineffective at stopping the German invasion because it failed to consider various potential modes of attack.

Security managers sometimes slap additional security devices, technologies, strategies, or measures onto their security to try to achieve "security in depth". The hope is that the adversary will have to defeat all these in sequence, making his job more difficult. Layered security can be quite expensive. More seriously, these new features can often be completely bypassed, or else they share a common mode of failure, or get in each other's way. Don't deploy additional security devices, technologies, strategies, or measures unless you thoroughly understand them, as well as what specific vulnerabilities in the original security you are trying to address.
     -- Roger Johnston

I think that the film *Clueless* was very deep. I think it was deep in the way that it was very light. I think lightness has to come from a very deep place if it's true lightness.
-- Actress Alicia Silverstone

18 - Homeland Security

Some Key Points:

• **Control is not security.**

• **Looking busy is not security.**

• **Being silly is not security.**

• **Trampling on the Constitution is not security.**

Regarding the color-coded terrorist warning system: "People focus too much on colors. It could be numbers. It could be animals."
    -- Homeland Security Director, Tom Ridge. [*Author's Comment:* How about Chicken, Opossum, Ostrich, Weasel, & Lemming?]

I thought the homeland security color codes were what color you were supposed to wear so that we could figure out who was the enemy. They should just get rid of the color code system and replace it with three levels: "Jesus Christ!", "God Damn It!", and "Fuck Me!"
    -- Louis Black

I'm not afraid of Terrorists…I have teenagers.
    -- Bumper sticker

Whoever fights monsters should see to it that in the process he does not become a monster.
    -- Friedrich Nietzsche (1844-1900)

We cannot deter if we cannot model the adversary.
    -- Richard Johnson

It's not winning if the good guys have to adopt the unenlightened, illegal, or morally reprehensible tactics of the bad guys.
    -- Roger Johnston

It is a question whether, when we break a murderer on the wheel, we do not fall into the error a child makes when it hits the chair it has bumped into.
    -- Georg C. Lichtenberg (1742-1799)

They who would give up an essential liberty for temporary security, deserve neither liberty or security.
    -- Benjamin Franklin (1706-1790)

I'm embarrassed for our country when I see how rudely and arrogantly foreign visitors are treated by Customs and Immigration at the airport. That's not what this country is all about.
    -- Anonymous

An intelligence service is, in fact, a stupidity service.
    -- E.B. White (1899-1985)

I believe that security declines as security machinery expands.
    -- E.B. White (1899-1985)

Our enemies are innovative and resourceful, and so are we. They never stop thinking about new ways to harm our country and our people, and neither do we.
    -- George W. Bush

Usually, terrible things that are done with the excuse that progress requires them are not really progress at all, but just terrible things.
    -- Russell Baker

Captain Blackadder: So in the name of security, sir, everyone who enters the room has to have his bottom fondled by this drooling pervert?
    -- From *Blackadder Goes Forth* (1989)

We have to get it right every day and the terrorists only have to get it right once. So we have to be ahead of the game.
    -- TSA Spokesperson

Reason by degrees submits to absurdity, as the eye is in time accommodated to darkness.
    -- Samuel Johnson (1709-1784)

Education is the cheapest defense of a nation.
    -- Edmund Burke (1729-1797)

National Monuments that a Department of Homeland Security memo listed as likely targets of terrorist attacks, and the funding they got to prevent them:
The Giant Lava Lamp in Soap Lake, Washington ($143 million)
Carhenge in Alliance, Nebraska ($25 million)
The World's Largest Ball of Paint in Alexandria, Indiana ($12 million)
The Museum of Bad Art in Dedham, Massachusetts ($31 million)
Barney Smith's Toilet Seat Art Museum in Alamo Heights, Texas ($10 million)

A large yellow sign spelling out "United States" on the U.S. side of the border facing Canada was removed from the new border station at Massena, NY because of security concerns. According to officials, "The sign could be a huge target and attract undue attention. Anything that would place our officers at risk we need to avoid."

Now that the world is getting over the initial shock, and the war against terrorism has begun, what now for bridal retailers?
    -- Actual 2002 editorial in the trade magazine *Bridal Buyer*

Like I said, desperate measures call for desperate times.
    -- Football player Jordan Babineaux

FBI Director J. Edgar Hoover liked to write comments in the margins of memos. One day he wrote, "Watch the Borders!" Nobody at FBI headquarters had the courage to question this order, so border security was increased at the Canadian and Mexican borders. It took a week before someone realized that Hoover was actually complaining about the small size of the memo's margins.

More warnings issued by all branches of the government today that another terrorist attack is imminent. We're not sure when, we're not sure where, just that it is coming. Who is attacking us now, the cable company?
    -- Jay Leno

Fear of a name increases fear of the thing itself.
    -- J.K. Rowling, *Harry Potter and the Sorcerer's Stone* (1997)

The core of our American democracy is the right to vote. Implicit in that right is the notion that that vote be private, that vote be secure, and that vote be counted as it was intended when it was cast by the voter. And I think what we're encountering is a pivotal moment in our democracy where all of that is being called into question.

-- Kevin Shelley

The Jones County, Mississippi election clerk who asked for, and received permission from, the county Board of Supervisors to remove the so-called "paper trail" printers from his county's 100% unverifiable Diebold touch-screen voting machines because, as he told the board reportedly (with a straight face), "The voting machines record every vote and there is no way for them to be tampered with."

I want to be buried in Louisiana, so I can stay active in politics.
  -- Governor Earl Long (1895-1960)

We will bankrupt ourselves in the vain search for absolute security.
  -- Dwight D. Eisenhower (1890-1969)

Insisting on perfect safety is for people who don't have the balls to live in the real world.
  -- Mary Shafer

There are some legitimate security issues, but I believe many of the objections the administration is making are not for security reasons, but to disguise mistakes that were made prior to September 11.
  -- Senator Bob Graham

Terrorists are nothing if not adaptable. We need to stop thinking like us, and start thinking like them.
  -- Steve Dahl

Rather than coming up with sensible security measures, every new measure is reactionary without any thought as to whether it works.
  -- Steven Potter

America can always be counted on to do the right thing—after it has exhausted all other possibilities.
  -- Winston Churchill (1874-1965)

Power is not a means, it is an end. One does not establish a dictatorship in order to safeguard a revolution; one makes the revolution in order to establish the dictatorship. The object of persecution is persecution. The object of torture is torture. The object of power is power.
  -- George Orwell (1903-1950)

I have deep concerns that a delicate and subtle shading/skewing of facts by you and others at the highest levels of FBI management has occurred and is occurring. ...I base my concerns on my relatively small, peripheral but unique role in the Moussaoui investigation in the Minneapolis Division prior to, during and after September 11th and my analysis of the comments I have heard both inside the FBI (originating, I believe, from you and other high levels of management) as well as your Congressional testimony and public comments. I feel that certain facts...have, up to now, been omitted, downplayed, glossed over and/or mis-characterized in an effort to avoid or minimize personal and/or institutional embarrassment on the part of the FBI and/or perhaps even for improper political reasons.

-- May 2002 letter from FBI Agent Coleen Rowley to FBI Director Robert Mueller concerning the FBI's handling of accusations that its inaction allowed the 9/11 attacks to happen

We cannot secure liberty and guarantee security simply by spending more and more money in the name of security. Every dollar misspent in the name of security weakens our already precarious economic condition, indebts us to foreign nations, and shackles the future of our children and grandchildren... We can only defend our freedoms by ensuring the dollars we spend on security are done so in a fiscally responsible manner, meet real needs, and respect the very rights we are aiming to preserve and protect.

-- Sen Thomas Coburn

Security is like liberty in that many are the crimes that are committed in its name.

-- Robert H. Jackson (1892-1954), dissenting opinion in *U.S. vs Shaughnessy*, 1950

19 - Privacy & Civil Liberties

Some Key Points:

• **Freedom is what we ultimately need to secure. It's counterproductive to harm freedom in the process of trying to protect it.**

Congress shall make no law respecting an establishment of religion, or prohibiting the free exercise thereof; or abridging the freedom of speech, or of the press; or the right of the people peaceably to assemble, and to petition the Government for a redress of grievances.
    -- 1st Amendment to the United States Constitution

The right of the people to be secure in their persons, houses, papers, and effects, against unreasonable searches and seizures, shall not be violated, and no Warrants shall issue, but upon probable cause, supported by Oath or affirmation, and particularly describing the place to be searched, and the persons or things to be seized.
    -- 4th Amendment to the United States Constitution

He that would make his own liberty secure must guard even his enemy from oppression; for if he violates his duty he establishes a precedent that will reach to himself.
    -- Thomas Paine (1737-1809)

Whenever a separation is made between liberty and justice, neither, in my opinion, is safe.
    -- Edmund Burke (1729-1797)

Living our values makes us stronger.
    -- Barak Obama

If we don't believe in freedom of expression for people we despise, we don't believe in it at all.
    -- Noam Chomsk

Governments who try to set up a situation where citizens think they must choose

between a free press and security are making a mistake that will ultimately weaken them, not strengthen them. It's not a real choice. It is a false choice."
   -- Gary Pruitt

Freedom is not a luxury that we can indulge in when at last we have security and prosperity and enlightenment; it is, rather, antecedent to all of these, for without it we can have neither security nor prosperity nor enlightenment.
   -- Henry Steele Commager (1902-1998)

Experience should teach us to be most on our guard to protect liberty when the Government's purposes are beneficent. Men born to freedom are naturally alert to repel invasion of their liberty by evil-minded rulers. The greatest dangers to liberty lurk in insidious encroachments by men of zeal, well-meaning but without understanding.
   -- Justice Louis D. Brandeis (1856-1941), dissenting, *Olmstead v. United States*, 277 US 479 (1928)

Free speech is intended to protect the controversial and even outrageous word; and not just comforting platitudes too mundane to need protection.
   -- Colin Powell

A patriot must always be ready to defend his country against its government.
   -- Edward Abbey (1927-1989)

It is the duty of every patriot to protect his country from its government.
   -- Thomas Paine (1737-1809)

Get a good night's sleep and don't bug anybody without asking me.
   -- Richard M. Nixon, 37th US President, to re-election campaign manager Clark MacGregor, recorded on tape later made public

"A matter of internal security"—the age-old cry of the oppressor.
   -- *Star Trek* Captain Jean-Luc Picard

Without censorship, things can get terribly confused in the public mind.
   -- General William Westmoreland (1914-2005)

I believe in censorship. I made a fortune out of it.
   -- Mae West (1893-1980)

You can protect your liberties in this world only by protecting the other man's freedom. You can be free only if I am free.
    -- Clarence Darrow (1857-1938)

The time has arrived...to get out of the shadow of states rights and walk forthrightly into the bright sunshine of human rights.
    -- Hubert H. Humphrey (1911-1978), from his renowned speech at the 1948 Democratic National Convention in Philadelphia

Toward no crime have men shown themselves so cold-bloodedly cruel as in punishing differences of belief.
    -- James Russell Lowell (1819-1891)

The greater the number of laws and enactments, the more thieves and robbers there will be.
    -- Lao-tzu (604 BC - 531 BC)

I disapprove of what you say, but I will defend to the death your right to say it.
    -- Attributed to Voltaire (1694-1778)

The will of the people is the only legitimate foundation of any government, and to protect its free expression should be our first object.
    -- Thomas Jefferson (1743-1826)

Patriotism is the last refuge of a scoundrel.
    -- Attributed to Samuel Johnson (1709-1784)

Most people want security in this world, not liberty.
    -- H.L. Mencken (1880-1956)

We must plan for freedom, and not only for security, if for no other reason than that freedom can make security secure.
    -- Karl Popper (1902-1994)

Power always has to be kept in check; power exercised in secret, especially under the cloak of national security, is doubly dangerous.
    -- William Proxmire (1915-2005)

The only thing that saves us from the bureaucracy is inefficiency. An efficient bureaucracy is the greatest threat to liberty.
    -- Eugene McCarthy (1916-2005)

Liberty means responsibility. That is why most men dread it.
-- George Bernard Shaw (1856-1950)

Better a thousandfold abuse of free speech than denial of free speech.
-- Charles Bradlaugh (1833-1891)

God grants liberty only to those who love it and are always ready to guard and defend it.
-- Daniel Webster (1782-1852)

If you can't beat them, arrange to have them beaten.
-- George Carlin (1937-2008)

The tree of liberty must be refreshed from time to time with the blood of patriots and tyrants.
-- Thomas Jefferson (1743–1826)

If liberty means anything at all, it means the right to tell people what they do not want to hear.
-- George Orwell (1903-1950)

Government is actually the worst failure of civilized man. There has never been a really good one, and even those that are most tolerable are arbitrary, cruel, grasping and unintelligent.
-- H.L. Mencken (1880-1956)

The policeman isn't there to create disorder, the policeman is there to preserve disorder.
-- Chicago Mayor Richard J. Daley responding to criticism of police tactics during the 1968 Democratic National Convention

May we never confuse honest dissent with disloyal subversion.
-- Dwight D. Eisenhower (1890-1969)

Relying on the government to protect your privacy is like asking a peeping tom to install your window blinds.
-- John Perry Barlow

Don't join the book burners. Do not think you are going to conceal thoughts by concealing evidence that they ever existed.

-- Dwight D. Eisenhower (1890-1969)

[Attorney General] Ashcroft went on to say that our way of life is being threatened by a group of radical religious fanatics who are armed and dangerous. And then he called for prayers in the schools and an end to gun control.
    -- Jay Leno

20 - Common Sense is Not All That Common/The Shallow End of the Gene Pool

Some Key Points:

• **You can't count on common sense, and shouldn't be surprised when you find it lacking.**

I would dance and be merry, life would be a ding-a-berry, if I only had a brain.
-- The Scarecrow, *The Wizard of Oz* (1939)

The problem with common sense is that it is not all that common.
-- Voltaire (1694-1778)

Common sense is as rare as genius.
-- Ralph Waldo Emerson (1803-1882)

Common sense is in spite of, not the result of, education.
-- Victor Hugo (1802-1885)

Against stupidity, the gods themselves struggle.
-- Doris Fleeson (1901-1970)

It takes a smart man to know he's stupid.
-- Barney Rubble, *The Flintstones*

We are all born ignorant, but one must work hard to remain stupid.
-- Benjamin Franklin (1706-1790)

The problem with the gene pool is that there is no lifeguard.
-- Steven Wright

Actual courtroom testimony:
Q: ALL your responses MUST be oral, OK? What school did you go to?
A: Oral.

Actual Tabloid Headlines:
•BIGFOOT CURED MY ARTHRITIS!

•Wonder drug makes people bulletproof!
•ABE LINCOLN WAS THE FATHER OF PROFESSIONAL WRESTLING!
•2,000-Year-Old Man Found in Tree…Wearing Watch That Still Ticks
•3 out of 5 Americans are reincarnated in Brazil
•MAN REINCARNATED AS HIMSELF!
•BEWARE! Squirrel brains can kill you
•PIZZA WAS SERVED AT THE LAST SUPPER

The doctors X-rayed my head and found nothing
     -- Baseball player Dizzy Dean after being hit on the head with a ball during the
     1934 World Series

Actual 1994 TV game show: *Trashed*. The premise: Contestants compete to answer
questions. The winning team gets to smash the losing team's prize possessions with
a sledgehammer.

At what elevation does an elk become a moose?
     -- Actual question asked of a Canadian park ranger

When did you build the glaciers?
     -- Actual question asked of park rangers at the Banff National Park in Canada

I live in a semi-rural area. We recently had a new neighbor call the local township
administrative office to request the removal of the Deer Crossing sign on our road.
The reason: many deer were being hit by cars and he no longer wanted them to
cross there.
     -- Anonymous

Following a trip to Greece, basketball player Shaquille O'Neal was asked whether he
had visited the Parthenon. "I can't really remember," Shaq replied, "the names of
the clubs we went to."

Our offense is like the Pythagorean theorem: There is no answer!
     -- Basketball player, Shaquille O'Neal

The world is made for people who aren't cursed with self-awareness.
     -- Annie Savoy (Susan Surandon), from the movie, *Bull Durham* (1988)

Actual courtroom testimony:
Lawyer: What about the research?

Witness: I don't think there is any research on that. There's a logical hunch that may be true, but I know of no research study that would support that.
Lawyer: What about common sense?
Witness: Well, I'm not here using common sense. I'm here as an expert.

I have plenty of common sense! I just choose to ignore it.
    -- Bill Watterson, from *Calvin and Hobbes*

Solutions are not the answer.
    -- Richard Nixon (1913-1994)

Factoid: One of the most common internet searches in late April and early May is people trying to find out what date Cinco de Mayo falls on that year.

Weakest Link host Anne Robinson: Watling Street, which now forms part of the A5, was built by which ancient civilization?
Contestant: Apes?

Game Show Host Eamonn Holmes: What travels at three hundred million miles a second?
Contestant: A cheetah.

If everyone wore my clothes, I don't think there would be wars, truly. Of course, then I would be the richest man in the world and most people would become bankrupt. My clothes are expensive. So maybe wars are better.
    -- Fashion designer Yves Saint Laurent (1936-2008)

Wasn't Winston Churchill the first black President of America? There's a statue of him near me that's black.
    -- British model Danielle Lloyd

I was recently on a tour of Latin America, and the only regret I have was that I didn't study Latin harder in school so I could converse with those people.
    -- Dan Quayle

The trouble with officials is that they just don't care who wins.
    -- Basketball coach Tommy Canterbury

When asked why he thought he had been reincarnated from earlier lives, British singer Lee Ryan said the proof was that "every time I each chicken, I eat it with my hands...like they did in the olden days."

The (UK) *Financial Times* was fooled by a 1998 April Fool press release claiming that Greenwich Mean Time would be rebranded as "Guinness Mean Time".

According to a 2010 Zogby poll, 37% of Americans believe in ghosts and 23% claim to have been visited by one.

I don't know if I'll like grits. Can I have just one?
   -- Customer to a truck stop waitress

Actual 911 recorded call:
Caller: Is it all right in Boulder to ask a woman for sex?

I was taking a bath in a Leningrad hotel when the floor concierge yelled that she had a cable for me. "Put it under the door," I cried. "I can't," she shouted. "It's on a tray!"
   -- Anthony Burgess

Narrator: As the 21st century began, human evolution was at a turning point. Natural selection, the process by which the strongest, the smartest, the fastest, reproduced in greater numbers than the rest, a process which had once favored the noblest traits of man, now began to favor different traits. Most science fiction of the day predicted a future that was more civilized and more intelligent. But as time went on, things seemed to be heading in the opposite direction. A dumbing down. How did this happen? Evolution does not necessarily reward intelligence. With no natural predators to thin the herd, it began to simply reward those who reproduced the most, and left the intelligent to become an endangered species.
   -- From the movie, *Idiocracy* (2006)

I do have a great relationship with God...He helped me when I had a drinking problem. He helped me through my divorce. He helped me through big decisions about the Knight Rider TV series.
   -- Actor David Hasselhoff

21 - Organizational Behavior, Bureaucracies, & Security Theater

Some Key Points:

• **Many of the attributes of large organizations and bureaucrats work against effective and efficient security.**

• **Beware of Security Theater.**

• **Protect Personally Identifiable Information (PII)!**

• **It is often necessary to be a rebel.**

Real Security makes you feel bad because you have to think and work hard, and because you will come to understand its problems, limitations, and vulnerabilities. Security Theater makes you feel good because it (falsely) purports to solve the problem relatively painlessly, without making you have to think.
-- Roger Johnston

The Official United States Department of Agriculture (USDA) Regulations for destroying old Woodsy Owl costumes: Incinerate the complete costume with the oversight of an official USDA Forest Service law enforcement officer. The entire Woodsy Owl costume, including each of the separate pieces, is to be destroyed beyond recognition. [If you do not have access to an official USDA Forest Service law enforcement representative, arrangements will be made for dealing with your costume by contacting the USDA-FS Washington Office at Woodsy Owl, c/o National Symbols Program, P.O. Box 96090, Washington, D.C. 20090-6090]

That's Entertainment Maxim: Ceremonial Security (a.k.a. "Security Theater") will usually be confused with Real Security; even when it is not, it will be favored over Real Security.

Fudd's Law: If you push on something hard enough, it will fall over.

If people don't want to come to the ballpark, how are you going to stop them?
-- Yogi Berra (1925-2015)

Electricity is really just organized lightning.

-- George Carlin (1937-2008)

You have enemies? Good. That means you've stood up for something sometime in your life.
-- Winston Churchill (1874-1965)

I hate to be a kicker,
I always long for peace,
But the wheel that does the squeaking,
Is the one that gets the grease.
-- Josh Billings (1818-1885)

When small men cast long shadows, the sun is going down.
-- Venita Cravens

Nothing will ever be attempted, if all possible objections must be first overcome.
-- Samuel Johnson (1709-1784)

Integrity is like oxygen. The higher you go, the less there is of it.
-- Paul Dickson

Those who concern themselves overmuch with little matters usually become incapable of dealing with larger ones.
-- François de La Rochefoucauld (1613-1680)

At some point in the project somebody will start whining about the need to determine the project "requirements". This involves interviewing people who don't know what they want but, curiously, know exactly when they need it.
-- Scott Adams

It is outrageous to know that security procedures are apparently so lax at the Department of Veterans Affairs that a single bureaucrat had the ability to put the personal information of over 26 million Veterans at risk for sale to the highest criminal bidder.
-- Bob Ney

Ad copy: "When a multivitamin was needed for an 11-year health study, the one chosen was Centrum® Silver®. One more reason to take the most recommended multivitamin brand." This is a remarkable sales pitch given that it occurred after results of the study were published, a study showing that Centrum Silver doesn't reduce the incidence of disease or mortality.

In October, after 18 months of investigation, the Texas State Library and Archives Commission concluded that the state government generated too many unnecessary and unread reports. The Commission issued its findings in a 668 page report.
-- *Minneapolis StarTribune*, Nov 22, 2007

"Your food stamps will be stopped effective March 1992 because we received notice that you passed away. May God bless you. You may reapply if there is a change in your circumstances."
-- Letter from the South Caroline Department of Social Services

Every man alone is sincere. At the entrance of a second person, hypocrisy begins. We parry and fend the approach of our fellow man by compliments, by gossip, by amusements, by affairs. We cover up our thought from him under a hundred folds.
-- Ralph Waldo Emerson (1803-1882)

"Career Alternative Enhancement Program"
-- Term used by Chrysler when laying off 5,000 employees

I've never seen a statue of a committee.
-- Anonymous

What is a committee? A group of the unwilling, picked from the unfit, to do the unnecessary.
-- Mark Twain (1835-1910)

A conference is a gathering of important people who singly can do nothing, but together can decide that nothing can be done.
-- Fred Allen (1894-1956)

What does it say about these two organizations that the FBI Headquarters is named after J. Edgar Hoover, a paranoid possible transvestite who had no respect for civil liberties or the constitution, while the Department of Energy headquarters is named after James Forrestal, the Secretary of Defense who had to be fired in 1949 by President Truman because of his paranoia and other psychological problems.
-- Anonymous

We're going to turn this team around 360 degrees.
-- Basketball play Jason Kidd

It is only doubt that creates. It is only the minority that counts.
-- H.L. Mencken (1880-1956)

Bucy's Law: Nothing is ever accomplished by a reasonable person.

The reasonable man adapts himself to the world; the unreasonable one persists in trying to adapt the world to himself. Therefore, all progress depends on the unreasonable man.
-- George Bernard Shaw (1856-1950)

The things we fear most in organizations—fluctuations, disturbances, imbalances—are the primary sources of creativity.
-- Margaret J. Wheatley

The status quo sucks.
-- George Carlin (1937-2008)

Status quo. Latin for "the mess we're in".
-- Ronald Reagan (1911-2004)

Bureaucracy defends the status quo long past the time when the quo has lost its status.
-- Laurence J. Peter (1919-1988)

Bureaucratic organizations by their very nature protect the venal and the incompetent because the purpose of a bureaucracy is to diffuse accountability so that no one person is ever responsible for the organization's failure.
-- Charles Hugh Smith

Behold the turtle. He makes progress only when he sticks his neck out.
-- James Bryant Conant (1893-1978)

Every society honors its live conformists and its dead troublemakers.
-- Mignon McLaughlin (1913-1983)

At least once in your lifetime, take a risk for a principle you believe in—even if it brings you up against your bosses.
-- Daniel Schorr (1916-2010)

Men, it has been well said, think in herds; it will be seen that they go mad in herds, while they only recover their senses slowly, and one by one.

-- Charles Mackay (1814-1889)

I believe in getting into hot water; it keeps you clean.
    -- Gilbert Keith Chesterton (1874-1936)

The hallmark of courage in our age of conformity is the capacity to stand on one's convictions—not obstinately or defiantly (these are gestures of defensiveness, not courage) nor as a gesture of retaliation—but simply because these are what one believes.
    -- Rollo May (1909-1994)

It's a good idea to obey all the rules when you're young just so you'll have the strength to break them when you're old.
    -- Mark Twain (1835-1910)

Disobedience, in the eyes of anyone who has read history, is man's original virtue. It is through disobedience and rebellion that progress has been made.
    -- Oscar Wilde (1854-1900)

Humanity has advanced, when it has advanced, not because it has been sober, responsible, and cautious, but because it has been playful, rebellious, and immature.
    -- Tom Robbins

Do what you feel in your heart to be right—for you'll be criticized anyway. You'll be damned if you do, and damned if you don't.
    -- Eleanor Roosevelt (1884-1962)

Stewart's Law: It is easier to get forgiveness than permission.

Show Me Maxim: No serious security vulnerability, including blatantly obvious ones, will be dealt with until there is overwhelming evidence and widespread recognition that adversaries have already catastrophically exploited it. In other words, "significant psychological (or literal) damage is required before any significant security changes will be made".

Hellrung's Law: If you wait long enough, it will go away.

Grelb's Law: But if it was bad, it will come back.

My husband once worked for a company that had a merit pay system. After six months they told him that he owed the company money.

-- Phyllis Diller

<u>Brien's First Law</u>: At some time in the life cycle of virtually every organization, its ability to succeed in spite of itself runs out.

A memorandum is written not to inform the reader but to protect the writer.
-- Dean Acheson (1893-1971)

Bureaucrats write memoranda both because they appear to be busy when they are writing and because the memos, once written, immediately become proof that they were busy.
-- Charles Peters

In a bureaucratic system, useless work drives out useful work.
-- Milton Friedman (1912-2006)

To get the attention of a large animal, be it an elephant or a bureaucracy, it helps to know what part of it feels pain. Be very sure, though, that you want its full attention.
-- Kelvin Throop

One of the most important ways to manifest integrity is to be loyal to those who are not present. In doing so, we build the trust of those who are present.
-- Stephen Covey

Hell, there are no rules here—we're trying to accomplish something.
-- Attributed to Thomas Edison (1847–1931)

We criticize the faults of others more out of pride than goodness; and we criticize them not so much to correct them as to persuade them that we are free from their faults.
-- François de La Rochefoucauld (1613-1680)

One man alone can be pretty dumb sometimes, but for real bona fide stupidity, there ain't nothin' can beat teamwork.
-- Edward Abbey (1927-1989)

Every organization is like a coconut tree full of monkeys. The ones at the top only see monkeys below them. The ones at the bottom only see assholes above them.
-- Anonymous

Is it ignorance or apathy? Hey, I don't know and I don't care.
  -- Jimmy Buffet

Just because I don't care doesn't mean I don't understand.
  -- Homer Simpson

I don't create controversies. They're there long before I open my mouth. I just bring them to your attention.
  -- Charles Barkley

Football incorporates the two worst elements of American society: violence punctuated by committee meetings.
  -- George Will

Actual title of a filing in the US. District Course, San Antonio:
DEFENDANT'S REPLY TO PLAINTIFF'S RESPONSE TO DEFENDANT'S MOTION TO STRIKE PLAINTIFF'S REPLY TO DEFENDANT'S OPPOSITION TO PLAINTIFF'S MOTION FOR PROTECTIVE ORDER, OR, IN THE ALTERNATIVE, APPLICATION FOR HEARING."

22 - Security Philosophy

Some Key Points:

• **Know thyself. Know thy weaknesses.**

• **Security is a compromise.**

• **"Displacement" (encouraging the bad guys to attack somebody else) can be useful, whatever the ethics of it.**

• **Honesty, imagination, diligence, and critical thinking are essential.**

"Bad guys attack and good guys react" is not a viable security strategy.
-- old proverb

We have only 2 modes of operation—complacency and pain.
-- James R. Schlesinger, Secretary of Energy

I like maxims that don't require behavior modification.
-- Bill Watterson, from *Calvin and Hobbes*

Newspapers are unable, seemingly, to discriminate between a bicycle accident and the collapse of civilization.
-- George Bernard Shaw (1856-1950)

To see what you are not is most important. Then what you are will naturally emerge.
-- Dick Olney

I believe in making the world safe for our children, but not our children's children, because I don't think children should be having sex.
-- Jack Handey

Trust not yourself, but your defects to know,
Make use of every friend and every foe.
-- Alexander Pope (1688-1774)

You can't depend on your judgment when your imagination is out of focus.
-- Mark Twain (1835-1910)

If you're tired of cardboard heroes saving the world in implausible ways, you're tired of life.
-- Anonymous

If everybody is thinking alike, then nobody is thinking.
-- George S. Patton (1885-1945)

All generalizations are false.
-- R.H. Grenier

General notions are generally wrong.
-- Lady Mary Wortley Montagu (1689-1762)

Distrust and caution are the parents of security.
-- Benjamin Franklin (1706-1790)

Don't let the best become the enemy of the good.
-- Anonymous

If you refuse to accept anything but the best, you often get it.
-- Anonymous

The pursuit of perfection often impedes improvement.
-- George Will

It is the greatest of all mistakes to do nothing because you can only do a little.
-- Sydney Smith (1771-1845)

If you can't be a positive example... maybe you can at least be a horrible warning to others.
-- Anonymous

It may be that your sole purpose in life is simply to serve as a warning to others.
-- Steven Wright

Some things are just too coincidental to be coincidental.
-- Yogi Berra (1925-2015)

There has never been a statue erected to the memory of someone who let well enough alone.
     -- Jules Ellinger

The only real security that a man can have in this world is a reserve of knowledge, experience, and ability.
     -- Henry Ford (1863-1947)

Ninety percent of everything is crap.
     -- Theodore Sturgeon (1918-1985)

Ninety percent of the game is half mental.
     -- Yogi Berra (1925-2015)

Only those means of security are good, are certain, are lasting that depend on yourself and your own vigor.
     -- Niccolo Machiavelli (1469-1527)

If at first you don't succeed, try again. Then quit. No use being a damn fool about it.
     -- W.C. Fields (1880-1946)

You don't have to swim faster than the shark, just faster than the guy next to you.
     -- Anonymous

In no direction that we turn do we find ease or comfort. If we are honest and if we have the will to win we find only danger, hard work and iron resolution.
     -- Wendell L. Willkie (1892-1944)

A man sits as many risks as he runs.
     -- Henry David Thoreau (1817-1862)

All great deeds and all great thoughts have a ridiculous beginning.
     -- Albert Camus (1913-1960)

Rules should be ways of thinking, not ways to avoid thinking.
     -- Anonymous

Security rules should increase security. They often don't.
     -- Roger Johnston

Advice to children crossing the street: Damn the lights! Watch the cars. The lights ain't never killed nobody.
　　-- Moms Mabley (1894-1975)

Those are my principles. If you don't like them, I have others.
　　-- Groucho Marx (1890-1977)

All things human hang by a slender thread; and that which seemed to stand strong suddenly fails and sinks in ruin.
　　-- Ovid (43 BC – 17 AD)

Being a blockhead is sometimes the best security against being cheated by a man of wit.
　　-- Francois La Rochefoucauld (1613-1680)

Believe nothing because a wise man said it,
Believe nothing because it is generally held,
Believe nothing because it is written,
Believe nothing because it is said to be divine,
Believe nothing because someone else believes it,
But believe only what you yourself judge to be true.
　　-- The Buddha (563? – 483? BC)

Buddhist to hotdog vendor: Make me One with everything.

The great enemy of truth is very often not the lie—deliberate, contrived, and dishonest—but the myth—persistent, persuasive, and unrealistic.
　　-- John F. Kennedy (1917-1963)

If you do not raise your eyes, you will think that you are the highest point.
　　-- Antonio Porchia (1885-1968)

There's such a thin line between winning and losing.
　　-- John R. Tunis (1889-1975)

The First Law of Thermodynamics: "You can't win."
The Second Law of Thermodynamics: "You can't break even."
The Third Law of Thermodynamics: "You can't quit."
　　-- Beat poet Allen Ginsberg (1926-1997)

The provision of effective security is paradoxically the first step towards decay, as an effective system will not only repel successful attacks, but also prevent the attacks being made … an illusion is then created that the established security is unnecessary suggesting decay will follow until the degree of security falls to the point where an attack will succeed.

-- Graham Underwood

Don't worry about the world coming to an end. It's always tomorrow in Australia.
-- Charles Schulz (1922-2000)

As to methods, there may be a million and then some, but principles are few. The man who grasps principles can successfully select his own methods. The man who tries methods, ignoring principles, is sure to have trouble.
-- Ralph Waldo Emerson (1803-1882)

23 - Security Planning & Strategy

Some Key Points:

- **Security arises from people, not from technology.**

- **Security rules that only the good guys follow may be pointless.**

- **There's a big difference between plans and planning.**

- **Seek simplicity and flexibility. Beware of complexity.**

- **Don't get so focused on prevention that you forget to plan for mitigation and resilience.**

- **Worry when things are going well.**

Strategy without tactics is the slowest route to victory. Tactics without strategy is the noise before defeat.
-- Sun Tzu (544 - 496 BC)

The silicon is fine. It's the carbon we have to deal with.
-- Mark Rasch after serious security breaches at the national laboratories

Why do they make me take off my cap and sunglasses at the bank? The only people who are going to honor that request are the ones who aren't there to rob the bank.
-- Anonymous

Plans are only good intentions unless they immediately degenerate into hard work.
-- Peter Drucker (1909-2005)

In preparing for battle, I have always found that plans are useless, but planning is indispensable.
-- Dwight D. Eisenhower (1890-1969)

It is a bad plan that admits of no modifications.
-- Publius Syrus (~42 BC)

It is not enough for a man to know how to ride. He must also know how to fall.
    -- Mexican Proverb

Hope is not a strategy.
    -- Michael Henos

If you're not confused, you're not thinking clearly.
    -- Irene Peter

Better to be despised for too anxious apprehensions, than ruined by too confident security.
    -- Edmund Burke (1729-1797)

When you know that you're capable of dealing with whatever comes, you have the only security the world has to offer.
    -- Harry Browne (1933-2006)

There is always an easy solution to every human problem—neat, plausible and wrong.
    -- H.L. Mencken (1880-1956)

The truth is rarely pure, and never simple.
    -- Oscar Wilde (1854-1900)

Seek simplicity, and distrust it.
    -- Alfred North Whitehead (1861-1947)

 Simplicity is the ultimate sophistication.
    -- Leonardo da Vinci (1452-1519)

Everything should be made as simple as possible, but not simpler.
    -- Albert Einstein (1879-1955)

One does not accumulate but eliminate. It is not daily increase but daily decrease. The height of cultivation always runs to simplicity.
    -- Bruce Lee (1940-1973)

I have yet to see any problem, however complicated, which, when you looked at it in the right way, did not become still more complicated.
    -- Poul Anderson (1926-2001)

A complex system that works is invariably found to have evolved from a simple system that works.
　　-- John Gaule

There are no differences but differences of degree between different degrees of difference and no difference.
　　-- William James (1842-1910), inspiration under nitrous oxide intoxication

We do not have enough strength to follow all the implications of our reasoning.
　　-- François de La Rochefoucauld (1613-1680)

[One way] researchers sometimes evaluate people's judgments is to compare those judgments with those of more mature or experienced individuals. This method has its limitations too, because mature or experienced individuals are sometimes so set in their ways that they can't properly evaluate new or unique conditions or adopt new approaches to solving problems.
　　-- Robert Epstein

In skating over thin ice, our safety is in our speed.
　　-- Ralph Waldo Emerson (1803-1882)

You mean they've scheduled Yom Kippur opposite Charlie's Angels?
　　-- Fred Silverman, TV programming executive, when told that Yom Kippur would fall on a Wednesday.

The current paradigm of 'protecting' infrastructure is unrealistic. We should shift our focus to that of resiliency.
　　-- James Carafano

Hawkin's Law: Progress does not consist of replacing a theory that is wrong with one that is right. It consists of replacing a theory that is wrong with one that is more subtly wrong.

An expert is a person who has made all the mistakes that can be made in a very narrow field.
　　-- Niels Bohr (1885-1962)

My definition of an expert in any field is a person who knows enough about what's really going on to be scared.
　　-- P.J. Plauger

As if there were safety in stupidity alone.
-- Henry David Thoreau (1817-1862)

There are risks and costs to a program of action. But they are far less than the long-range risks and costs of comfortable inaction.
-- John F. Kennedy (1917-1963)

There is a time when what you've creating and the environment you're creating it in come together.
-- Grace Harrtigan (1922-2008)

Research is what I am doing when I don't know what I am doing.
-- Wernher von Braun (1912–1977)

Be careful that victories do not carry the seed of future defeats.
-- Ralph W. Sockman (1889-1970)

You can only cure retail but you can prevent wholesale.
-- Brock Chisholm (1896-1971)

I never know what I think about something until I read what I've written on it.
-- William Faulkner (1897-1962)

"Compliance-Based Security" is an oxymoron.
-- Anonymous

By the Book Maxim: Full compliance with security rules and regulations is not compatible with optimal security.

I have not failed. I've just found 10,000 ways that won't work.
-- Attributed to Thomas A. Edison (1847–1931)

A man cannot be too careful in the choice of his enemies.
-- Oscar Wilde (1854-1900)

The opportunity for doing mischief is found a hundred times a day, and of doing good once in a year.
-- Voltaire (1694-1778)

Evil is easy, and has infinite forms.
-- Blaise Pascal (1623-1662)

Evil will always triumph because good is dumb.
-- Rick Moranis, as Dark Helmet in *Spaceballs* (1987)

The evil of the world is made possible by nothing but the sanction you give it.
-- Ayn Rand (1905-1982)

The time to repair the roof is when the sun is shining.
-- John F. Kennedy (1917-1963)

Human beings, who are almost unique in having the ability to learn from the experience of others, are also remarkable for their apparent disinclination to do so.
-- Douglas Adams (1952-2001)

How is it that a single match can start a giant forest fire, but you need a whole box of matches to light the barbeque?
-- Anonymous

I never expect to see a perfect work from imperfect man.
-- Alexander Hamilton (1755?-1804)

Ring the bells that still can ring.
Forget your perfect offering.
There is a crack in everything.
That's how the light gets in.
-- The song *Anthem*, by Leonard Cohen

We spend all our time searching for security, and then we hate it when we get it.
-- John Steinbeck (1902-1968)

The most erroneous stories are those we think we know best—and therefore never scrutinize or question.
-- Stephen Jay Gould (1941-2002)

There is no such thing as security. There never has been.
-- Germaine Greer

Ask for no guarantees, ask for no security, there never was such an animal. And if there were, it would be related to the great sloth which hangs upside down in a tree all day every day, sleeping its life away.
-- Ray Bradbury (1920-2012)

There is no security on this earth; there is only opportunity.
    -- Attributed to Douglas Macarthur (1880-1964)

Life is either a daring adventure or nothing. Security does not exist in nature, nor do the children of men as a whole experience it. Avoiding danger is no safer in the long run than exposure.
    -- Helen Keller (1880-1968)

Falling flat on your face is still moving forward.
    -- Anonymous

The only fence against the world is a thorough knowledge of it.
    -- John Locke (1632-1704)

A false sense of security is worse than no security at all, because you let your guard down.
    -- Anonymous

Whenever anyone says, 'theoretically', they really mean, 'not really'.
    -- Dave Parnas

What if there were no hypothetical situations?
    -- Anonymous

It was beautiful and simple as all truly great swindles are.
    -- O. Henry (William Sydney Porter) (1862-1910)

The hardness of the butter is directly proportional to the softness of the bread.
    -- Anonymous

When you are in any contest you should work as if there were—to the very last minute—a chance to lose it.
    -- Dwight D. Eisenhower (1890-1969)

The superior man, when resting in safety, does not forget that danger may come. When in a state of security he does not forget the possibility of ruin. When all is orderly, he does not forget that disorder may come. Thus his person is not endangered, and his States and all their clans are preserved.
    -- Confucius (551 – 479 BC)

He is safe from danger who is on guard even when safe

-- Publilius Syrus (~42 BC)

He that's secure is not safe.
   -- Benjamin Franklin (1706-1790)

And he that makes his soul his surety
I think does give the best security.
   -- Samuel Butler (1612-1680)

A ship in harbor is safe, but that is not what ships are built for.
   -- John A. Shedd

Opportunity makes the thief.
   -- English proverb

You have to be careful if you don't know where you are going because you might not get there.
   -- Yogi Berra (1925-2015)

Security is the essential roadblock to achieving the road map to peace.
   -- George W. Bush, Washington, D.C., July 25, 2003

If you see the world in black and white, you're missing important grey matter.
   -- Jack Fyock

It is easier for me to see everything as one thing than to see one thing as one thing.
   -- Antonio Porchia (1885-1968)

If the only tool you have is a hammer, you tend to see every problem as a nail.
   -- Abraham Maslow (1908-1970)

The odds of hitting your target go up dramatically when you aim at it.
   -- Mal Pancoast

I stuck to my game plan—stumbling forward and getting hit in the face.
   -- Boxer Randall "Tex" Cobb

Everybody has a plan until they get hit in the face.
   -- Variously attributed to boxers Mike Tyson, Leon Spinks, and Joe Louis (1914-1981), and also to Sun Tzu (544-496 BC)

There's more to boxing than hitting. There's not getting hit, for instance.
  -- Boxer George Foreman

We didn't lose the game. We just ran out of time.
  -- Vince Lombardi (1913-1970)

Security is mostly about what you get wrong, not what you get right.
  -- Roger Johnston

The chain of security is only as good as the weakest link.
  -- old adage

I never had a policy; I have just tried to do my very best each and every day.
  -- Abraham Lincoln (1809-1865)

24 - Security Management & Leadership

Some Key Points:

• **Management is what they hired you for; Leadership is what they need.**

• **"Bad Manager" is an oxymoron; "Enlightened Leadership" is a redundancy.**

• **Humor is useful and essential.**

Find the right employees, put them in the right system, and supervision is unnecessary...Every minute spent on management is wasted...The time spent managing employees is one measure of the failure of hiring and organizing.
 -- Dale Dauten

A good manager is best when people barely know that he exists. Not so good when people obey and acclaim him. Worse when they despise him.
 -- Lao-Tzu (604 BC – 531 BC)

For every hour of brains, you will be charged three hours. The other two hours go to management and project management, which is to say they are wasted.
 -- Robert Cringely

When my husband comes home, if the kids are still alive, I figure I've done my job.
 -- Roseanne Barr

I consider myself to be a pretty good judge of people...that's why I don't like any of them.
 -- Roseanne Barr

You don't realize how easy this game is until you get up in that broadcasting booth.
 -- Mickey Mantle (1931-1995)

The key to being a good manager is keeping the people who hate me away from those who are still undecided.
 -- Casey Stengel (1890-1975)

You do not lead by hitting people over the head. That's assault, not leadership.
    -- Dwight D. Eisenhower (1890-1969)

Church's new fund raising slogan: "I upped my pledge—up yours!"

If you have responsibility for security but have no authority to set rules or punish violators, your own role in the organization is to take the blame when something big goes wrong.
    -- Eugene Spafford

Just remember, there's a right way and a wrong way to do everything and the wrong way is to keep trying to make everybody else do it the right way.
    -- M*A*S*H, Colonel Potter

The world always makes the assumption that the exposure of an error is identical with the discovery of truth—that the error and truth are simply opposite. They are nothing of the sort. What the world turns to, when it is cured on one error, is usually simply another error, and maybe one worse than the first one.
    -- H.L. Mencken (1880-1956)

A sense of humor is part of the art of leadership, of getting along with people, of getting things done.
    -- Dwight D. Eisenhower (1890-1969)

We trained hard, but it seemed that every time we were beginning to form up into teams, we would be reorganized. I was to learn later in life that we tend to meet any new situation by reorganizing; and a wonderful method it can be for creating the illusion of progress while producing confusion, inefficiency, and demoralization.
    -- Petronius Arbiter, 210 BC

Definition—Chief Security Officer (CSO): The manager we marginalize, under-fund, and then blame for any security incidents.

Some of us learn from other people's mistakes. The rest of us are the other people.
    -- Anonymous

You can buy muscles, but you can't buy cojones.
    -- Bas Rutten

Dick Solomon: I'm sorry, there is simply no room in the budget for raises. But I can go you one better: promotions! Sally, you are now "Senior" Security Officer.
-- *Third Rock from the Sun*

"I just work here."
-- The common utterance/attitude of disengaged employees everywhere

"Tut, tut, child," said the Duchess. "Everything's got a moral if only you can find it."
-- Lewis Carroll (1832-1898)

I don't have the first clue who he is talking about because all I worry about is Jerome.
-- Basketball player Jerome James, responding to criticism from his coach that he was selfish

It is a puzzling thing. The truth knocks on the door and you say, go away, I'm looking for the truth. And so it goes away. Puzzling.
-- Robert M. Pirsig

Many of us have heard opportunity knocking at our door, but by the time we unhooked the chain, pushed back the bolt, turned two locks, and shut off the burglar alarm—it was gone.
-- Anonymous

Consider the postage stamp: its usefulness consists in the ability to stick to one thing till it gets there.
-- Josh Billings (1818-1885)

If a man does his best, what else is there?
-- George S. Patton (1885-1945)

If you aspire to the highest place, it is no disgrace to stop at the second, or even the third place.
-- Cicero (106 BC – 43 BC)

I have nothing against work, particularly when performed, quietly and unobtrusively, by someone else. I just don't happen to think it's an appropriate subject for an "ethic."
-- Barbara Ehrenreich

Don't be a piano player in a whorehouse.
    -- John S. Carroll

Learn to pause...or nothing worthwhile will catch up to you.
    -- Doug King

This project is so important, we can't let things that are more important interfere with it.
    -- UPS memo

Now, here, you see, it takes all the running you can do, to keep in the same place. If you want to get somewhere else, you must run at least twice as fast as that.
    -- Lewis Carroll (1832-1898), *Through the Looking-Glass*

Quote from the boss at Citrix Corporation: "Teamwork is a lot of people doing what I say."

I don't never have any trouble in regulating my own conduct, but to keep other folks' straight is what bothers me.
    -- Josh Billings (1818-1885)

If everything seems under control, you're just not going fast enough.
    -- Mario Andretti

If you don't win, you're going to be fired. If you do win, you've only put off the day you're going to be fired.
    -- Baseball manager Leo Durocher (1905-1991)

So much of what we call management consists in making it difficult for people to work.
    -- Peter Drucker (1909-2005)

Management is doing things right; leadership is doing the right things.
    -- Peter Drucker (1909-2005)

Any good orchestra can play a concert without a conductor, but name one conductor that can play a concert without the orchestra!
    -- Anonymous

A reporter in the 60's was walking around the halls of NASA one day and ran across a janitor. He asked the janitor what he was doing, and the janitor replied, "I'm helping put a man on the moon".
-- Anonymous

Skills are cheap; chemistry is expensive.
-- Mal Pancoast

The best morale exists when you never hear the word mentioned. When you hear a lot of talk about it, it is usually lousy.
-- Dwight D. Eisenhower (1890-1969)

Once the game's over, the King and the pawn go back in the same box.
-- Italian proverb

When a journalist asked if class barriers had been eliminated in England, author Barbara Cartland replied, "Of course they have or I wouldn't be sitting here talking to someone like you."

Leadership is based on inspiration, not domination; on cooperation, not intimidation.
-- William Arthur Wood

If you want to build a ship, don't drum up people together to collect wood and don't assign them tasks and work, but rather teach them to long for the endless immensity of the sea.
-- Antoine de Saint-Exupery (1900-1944)

Leadership is communicating to people their worth and potential so clearly that they come to see it in themselves.
-- Stephen Covey

Leadership is lifting a person's vision to higher sights, the raising of a person's performance to a higher standard.
-- Peter Drucker (1909-2005)

The leaven of true leadership cannot lift others unless we are with and serve those to be led.
-- Spencer W. Kimball (1895-1985)

To get the best out of a man, go to what is best in him.

-- Daniel Considine

One of the hardest tasks of leadership is understanding that you are not what you are, but what you're perceived to be by others.
  -- Edward L. Flom

Good leadership consists in showing average people how to do the work of superior people.
  -- John D. Rockefeller (1839-1937)

Leadership: The art of getting someone else to do something you want done because he wants to do it.
  -- Dwight D. Eisenhower (1890-1969)

The only thing worse than a man you can't control is a man you can.
  -- Margo Kaufman

The primary purpose of management is to kill any hope that staying in your current job will work out for you. Bad management is how imagination gets wings. The economy needs workers who are fed up, desperate, and willing to quit their jobs for something better. You can't do something great until first you quit something that isn't…The economy needs hamster-brained sociopaths in management to drive down the opportunity cost of entrepreneurship. Luckily, we're blessed with an ample supply.
  -- Scott Adams

Be who you are and say what you feel, because those who mind don't matter and those who matter don't mind.
  -- Dr. Seuss (Theodor Seuss Geisel) (1904-1991)

Power is always dangerous. Power attracts the worst and corrupts the best.
  -- Edward Abbey (1927-1989)

Nearly all men can stand adversity, but if you want to test a man's character, give him power.
  -- Abraham Lincoln (1809-1865)

Power tends to corrupt, and absolute power corrupts absolutely.
  -- Lord Acton (1834-1902)

Power corrupts. Absolute power is kind of neat.

-- John Lehman (1981-1987), Secretary of the Navy

The power to command frequently causes failure to think.
   -- Barbara W. Tuchman (1912-1989), *The March of Folly*

I like to buy a company any fool can manage because eventually one will.
   -- Mutual fund manager Peter S. Lynch

In a hierarchy every employee tends to rise to his level of incompetence.
   -- *The Peter Principle* (Laurence J. Peter and Raymond Hull)

Everybody is a genius. But, if you judge a fish by its ability to climb a tree, it will spend its whole life believing that it is stupid.
   -- Albert Einstein (1879-1955)

Men will follow this officer—if only out of sheer curiosity.
   -- From an military officer's annual performance appraisal

A scout troop consists of twelve little kids dressed like schmucks following a big schmuck dressed like a kid.
   -- Jack Benny (1894-1974)

To those of you who received honors, awards, and distinctions, I say, well done! And to the C students, I say, you too can be President of the United States.
   – George W. Bush, commencement address at Yale, 2001

25 - Security Practice

Some Key Points:

- **The Real-World is nowhere anybody should want to work.**

- **Practice is not Theory.**

- **Good security requires critical thinking.**

It's called "practice" because you're always hoping to get better at it.
-- Anonymous

Yogi Berra's Theory: In theory there is no difference between theory and practice. In practice there is.

Horngren's Law: The Real World is a special case.

Learn the principle, abide by the principle, dissolve the principle.
-- Bruce Lee (1940-1973)

If you aren't fired with enthusiasm, you will be fired with enthusiasm.
-- Vince Lombardi (1913-1970)

Sincerity is everything. If you can fake that, you've got it made.
-- George Burns (1885-1996)

It is dangerous to be sincere unless you are also stupid.
-- George Bernard Shaw (1856-1950)

Between the idea and the reality,
Between the motion
And the act
Falls the Shadow.
-- T.S. Eliot (1888-1965), *The Hollow Men*

Mediocrity is climbing molehills without sweating.
-- Icelandic proverb

<u>Tucker's Third Maxim</u>: If you're not failing when you're training or testing your security, you're not learning anything.

Haven't you heard the phrase, 'The customer is always right?'
Let me tell you something. Let me give you a little secret, okay? THE CUSTOMER IS ALWAYS AN ASSHOLE!
    -- Dialog from the movie, *Mallrats* (1995)

When you say that you agree with a thing in principle you mean that you have not the slightest intention of carrying it out in practice.
    -- Otto von Bismarck (1815-1898)

All methodologies are based on fear.
    -- Kent Beck

I have hunted deer on occasions, but they were not aware of it.
    -- Felix Gear

If a thing is worth doing, it is worth doing slowly . . . very slowly.
    -- Burlesque artist Gypsy Rose Lee (1911-1970)

Who's in charge of security? Probably the same guys who protected Sonny Corleone at the tollbooth.
    -- Billy Crystal at the 2000 Academy Awards

Don't play what's there, play what's not there.
    -- Miles Davis (1926-1991)

There's a fine line between fishing and just standing on the shore like an idiot.
    -- Steven Wright

Never interrupt your enemy when he is making a mistake.
    -- Napolean Bonaparte (1769-1821)

The only security is the constant practice of critical thinking.
    -- William Graham Sumner (1840-1910)

Delight at having understood a very abstract and obscure system leads most people to believe in the truth of what it demonstrates.
    -- Georg C. Lichtenberg (1742-1799)

We made too many wrong mistakes.
-- Yogi Berra (1925-2015)

How nice to have all the wrong things in one place.
-- Astronomer Arthur Eddington (1882-1944) about a book by his rival

I skate to where the puck is going to be, not where it has been.
-- Wayne Gretzky

It's a funny thing. The more I practice, the luckier I get.
-- Golfer Arnold Palmer

Half the game is mental; the other half is being mental.
-- Hockey player Jim McKenny

The depressing thing about tennis is that no matter how good I get, I'll never be as good as a wall.
-- Mitch Hedberg

Weakest Link Maxim: The efficacy of security is determined more by what is done wrong than by what is done right. (This is because the bad guys don't typically attack randomly, but rather at the weakest point.)

Just because nobody complains doesn't mean that all parachutes are perfect.
-- Benny Hill (1924-1992)

26 - Security Guards

Some Key Points:

• **We put our frontline personnel on the front lines, so empower and motivate them to do a good job.**

• **Excellence and character come with practice.**

• **Security is mostly about paying attention, but paying attention is hard.**

• **The ear is better at detecting something anomalous than the eye.**

But who is to guard the guards themselves?
   -- Juvenal (55 – 127 AD)
Plato (~427 - ~347 BC) offered one possible answer: They will guard themselves against themselves. We must tell the guardians a noble lie. The noble lie will inform them that they are better than those they serve and it is therefore their responsibility to guard and protect those lesser than themselves. We will instill in them a distaste for power or privilege; they will rule because they believe it right, not because they desire it.

Harry Solomon: I didn't have enough experience to sell hot dogs, so they hired me as a security guard.
   -- From the television show, *Third Rock from the Sun* (1996)

Hey, I'm a security officer! I took a 2 week course.
   -- Dialog from the movie, *Sleepover* (2004)

-So what made you want to pursue security?
-I never finished high school. This is all I could get.
   -- Dialog from the movie, *Paul Blart: Mall Cop* (2009)

Definition—closed circuit television (CCTV): A video system designated as "closed circuit" to distinguish it from broadcast TV, which the guys at the guard station are really watching.

Security Guard: Where's the TV guide?

-- Last line in the , *The Truman Show* (1998)

Security Guard: Don't make me take off my sunglasses!
    -- From the movie, *Bringing Out the Dead* (1999)

Definition—proprietary guard force: Our guards view their job as some kind of entitlement.

Definition—contract guard force: Now we can blame our lousy security on somebody else.

I don't need any bodyguards.
    -- Jimmy Hoffa (1913-1975)

Rachel: Well, you don't look like a bodyguard.
Frank: What'd you expect?
Rachel: Well, I don't know, maybe a tough guy?
Frank: This is my disguise.
    -- From the movie, *The Bodyguard* (1992)

Factoid: Secret Service agents wear sunglasses not just to look cool and intimidating, but because it disguises where they are looking.

Singer-songwriter Bob Dylan was once stopped by his own security guards at a show. He said, "I can hardly blame them. Look at me!"

Budapest zoo sign in English: Please do not feed the animals, if you have any suitable food, give it to the guard on duty.

Excellence is an art won by training and habituation. We do not act rightly because we have virtue or excellence, but we rather have those because we have acted rightly. We are what we repeatedly do. Excellence, then, is not an act but a habit.
    -- Aristotle (384 – 322 BC)

Habits change into character.
    -- Ovid (43 BC – 17 AD)

A hypothetical paradox: what would happen in a battle between an Enterprise security team, who always get killed soon after appearing, and a squad of Imperial Stormtroopers, who can't hit the broad side of a planet?
    -- Tom Galloway

Never beam down in a red shirt (the traditional uniform color of security personnel on Star Trek). [You won't be beaming back up.]
   -- Robert W. Bly

Once you have missed the first buttonhole, you'll never manage to button up.
   -- Johann Wolfgang von Goethe (1749-1832)

How difficult it is not to betray one's guilt by one's looks.
   -- Ovid (43 BC – 17 AD)

The spirited horse, which will try to win the race of its own accord, will run even faster if encouraged.
   -- Ovid (43 BC – 17 AD)

It is not enough to do your best; you must know what to do, and THEN do your best.
   -- W. Edwards Deming (1900-1993)

Only a mediocre person is always at his best.
   -- W. Somerset Maugham (1874-1965)

The simple act of paying attention can take you a long way.
   -- Keanu Reeves

Caffeine Maxim: On a day-to-day basis, security is mostly about paying attention.

Any Donuts Left? Maxim: But paying attention is very difficult.

Factoid: Psychologists know that observers spend decreasing amounts of time and effort carefully looking for rare events when they haven't occurred in a while.

If you don't find it often, you often don't find it.
   -- Jeremy M. Wolfe

Factoid: "Perceptual Blindness" is a problem for all human beings, not just security guards. People tend to see what they expect to see and what they are prepared to see, not necessarily what is really there.

We don't see things as they are, we see things as we are.
   -- Anais Nin (1903-1977)

What we see depends mainly on what we look for.
    -- John Lubbock (1834-1913)

It all depends on how we look at things, and not on how things are in themselves.
    -- Carl Jung (1875-1961)

It's better to be looked over than overlooked.
    -- Mae West (1893-1980), *Belle of the Nineties*, 1934

The ear tends to be lazy, craves the familiar and is shocked by the unexpected; the eye, on the other hand, tends to be impatient, craves the novel, and is bored by repetition.
    -- W.H. Auden (1907-1973)

If somebody shows up wearing a Fire Marshall's hat and claiming it's an emergency, are you going to let him in?

In moments of crisis, one is never fighting against an external enemy but always against one's own body.
    -- George Orwell (1903-1950), *1984*

Exceeded expectations? I'd say he's done more than that!
    -- Yogi Berra (1925-2015)

If you need 2 yards, I'll get you 3. If you need 10 yards, I'll get you 3.
    -- Running back Leroy Hoard

The only qualifications for a lineman are to be big and dumb. To be a back, you only have to be dumb.
    -- Knute Rockne (1888-1931)

If you see a whole thing, it seems that it's always beautiful. Planets, lives... But up close, a world's all dirt and rocks. And day to day, life's a hard job, you get tired, you lose the pattern.
    -- Ursula Le Guin (1929-2018)

27 - Communication

Some Key Points:

- **Communication is job 1.**

Actual overheard conversation between two teenage girls:
--So he's like, 'nuh uh,' and I'm like, 'uh huh,' and he's like, 'nuh uh,' and I'm like, 'um…uh huh,' and he's like, 'nuh uh.'
--No way!
--Way.

We know that communication is a problem, but the company is not going to discuss it with the employees.
    -- Infamous AT&T memo

I hope you believe you understand what you think I said, but I'm not sure you realize that what you've heard is not what I meant.
    -- Richard Nixon (1913-1994)

I'll phone up [the Hilton hotel in Paris] and say, "Hi, it's Paris Hilton," and they'll say, "Yes, this is the Paris Hilton." So I'm like, "Yes, I know, I'm Paris Hilton." It can go on for hours like some bad comedy film.
    -- Paris Hilton

If you would win a man to your cause, first convince him that you are his sincere friend.
    -- Abraham Lincoln (1809-1865)

I've been laboring here for five years and now we have a sock talking at our commencement.
    -- Recent college graduate Samantha Chie on her college's decision to give an honorary degree to Kermit the Frog

When I was in high school, I got in trouble with my girlfriend's father. He said, "I want my daughter back by 8:15." I said, "The middle of August? Cool!"
    -- Steven Wright

British transportation officials in 2008 sent an e-mail to a well-known Welsh Councilman asking for a translation of a road sign that was intended to say, "No entry for heavy goods vehicles." When they got back a reply, they dutifully put up a road sign that said in Welsh, "I am not in the office at the moment."

My wife keeps complaining I never listen to her... or something like that.
-- Bumper sticker

I type at 101 words a minute, but it's not in my own language.
-- Mitch Hedberg

It's not so difficult to communicate effectively with the other gender if you just have a little patience and keep in mind that they are often irrational, stubborn, inconsiderate, self-absorbed, and can be remarkably dim-witted at times. Both men and women seem to appreciate this advice.
-- Anonymous

Dear Mr. Cook:
We have attempted on several occasions to reach you by telephone to discuss payment of your telephone account—which was recently disconnected.
-- Actual letter from the phone company

I have a phone answering machine that uses synthesized speech to announce the Caller ID. Unfortunately, the manufacturer never taught it that "ph" is the same as the "f" sound. The machine frequently announces that the caller is a "cell puh-hone". It's a little disconcerting that a phone accessory that mimics human speech doesn't have enough self-awareness to know how to pronounce "phone".
-- Roger Johnston

What we got here is a failure to communicate
-- From the movie, *Cool Hand Luke* (1967)

Harry Potter this ... Harry Potter that ... I'd never even heard of Harry Potter until the book came out!
-- Caller, BBCRadio 5 Live

Radio Communications Log:
Captain of aircraft carrier: Please divert your course 0.5 degrees to the south to avoid a collision.

Canadian reply: Recommend you divert your course 15 degrees to the south to avoid a collision.
Captain of aircraft carrier: This is the captain of a U.S. Navy ship. I say again, divert your course!
Canadian reply: No. I say again, you divert YOUR course!
Captain of aircraft carrier: THIS IS THE AIRCRAFT CARRIER USS CORAL SEA. WE ARE A LARGE WARSHIP OF THE U.S. NAVY. DIVERT YOUR COURSE NOW!!
Canadian reply: This is a lighthouse. Your call.
   -- Actual radio exchange off the coast of Newfoundland, October 10, 1995

Advice on writing:
Remember to never split an infinitive. Take the bull by the hand and avoid mixed metaphors. Proofread carefully to see if you works out. Avoid clichés like the plague. And don't overuse exclamation marks!
   -- William Safire (1929-2009)

Actual Tech Support phone conversation:
Tech support: What does the screen say now?
Customer: It says, "Hit ENTER when ready."
Tech support: Well?
Customer: How do I know when it's ready?

CNN anchor Wolf Blitzer: Does that mean you want to come up with a new Sarah Palin initiative that you want to release right now?
Alaska Governor Sarah Palin: Gah! Nothing specific right now. Sitting here in these chairs that I'm going to be proposing but in working with these governors who again on the front lines are forced to and it's our privileged obligation to find solutions to the challenges facing our own states every day being held accountable, not being just one of many just casting votes or voting present every once in a while, we don't get away with that.

The boat in the 1960's television show *Gilligan's Island* was named the *S.S. Minnow* to make fun of Newton Minow. He was the Chairman of the Federal Communications Commission who made the famous accusation that television was a "vast wasteland".

George: Gracie, where did you get the flowers?
Gracie: I went to visit Mabel in the hospital.
George: Ok, but where did you get the flowers?

Gracie: Well, you told me to take her flowers.
George: Say goodnight, Gracie.
Gracie: Goodnight, Gracie.
   -- George Burns (1896-1996) and Gracie Allen (1895? – 1964)

News Correction: In a news story Friday ("Spectrum Holds Condom Olympics to Educate on Safe Sex," page 3), it was incorrectly stated due to a reporting error that health and wellness educator Beth Grampetro and Tim Hegan, an ORL area director, said Fruit Roll-Ups are adequate protection against STDs. No health officials said or advocated this use at the Condom Olympics.
   -- *The Daily Free Press*

According to the United Nations, the most common reason for condoms being ineffective in developing countries is that men wear condoms on their finger. The most common pitfall with the pill is that men take it instead of women.

Actual 911 recorded conversation:
911: 9-1-1. What's the nature of your emergency, please?
Caller: I'm trying to reach nine-eleven, but my phone doesn't have an eleven on it.
911: This is nine-eleven.
Caller: I thought you said it was nine-one-one.
911: Yes, ma'am. Nine-one-one and nine-eleven are the same thing.
Caller: Honey, I may be old, but I'm not stupid.

What's the point of Esperanto? There's already an international language. It's called Bad English.
   -- Theodore von Karman (1881-1963)

28 - Security Training

Some Key Points:

• **Training needs to be education, not pronouncement.**

• **It's got to be made interesting, relevant, and motivational, and not just consist of threats and lists of no-nos.**

Definition—Security Awareness Training: Presentations that convince employees who once vaguely believed that security was a good idea that they were sadly mistaken.

Definition—Counter-Intelligence Program: Security Awareness Training that is so awful, it insults everyone's intelligence.

I'm such a good lover because I practice a lot on my own.
    -- Woody Allen

Karate is a form of martial arts in which people who have had years and years of training can, using only their hands and feet, make some of the worst movies in the history of the world.
    -- Dave Barry

In my organization, our online training course in Ethics is password protected. This is presumably to prevent unauthorized personnel from stealing our unique ideas on ethical behavior.
    -- Roger Johnston

A report by Ernst & Young finds that, "Security awareness programs at many organizations are weak, half-hearted and ineffectual." As a result, employees ignore them.

We don't train people, we train dogs. We educate, motivate, and remind people.
    -- Anonymous

People need to be reminded more often than they need to be instructed.
    -- Samuel Johnson (1709-1784)

Most managers were trained to be the thing they most despise—bureaucrats.
   -- Alvin Toffler

Don't let school get in the way of your education.
   -- Mark Twain (1835-1910)

Education's purpose is to replace an empty mind with an open one.
   -- Malcolm Forbes (1919-1990)

Education is not the filling of a pail, but the lighting of a fire.
   -- William Yeats (1865-1939)

I hear and I forget. I see and I remember. I do and I understand.
   -- Confucius (551 BC – 479 BC)

Attention to a subject depends upon our interest in it. We rarely forget that which
has made a deep impression on our minds.
   -- Tryon Edwards (1809-1894)

Boxing is like jazz. The better it is, the less people appreciate it.
   -- George Foreman

Riley Hale: You know, these exercises are fantastic. When the day comes that we
have to go to war against Utah, we're really gonna kick ass, y'know?"
   -- From the movie, *Broken Arrow* (1996)

29 - Security Metrics & Standards

Some Key Points:

• **How do you know how well you're doing if you don't measure?**

• **But knowing what to measure and how is a challenge.**

• **Standards aren't always helpful.**

• **Beware the Fallacy of Precision: Just because you can arbitrarily assign a number to something (e.g., a threat probability) does not mean you understand that thing or that the number has any real meaning.**

Thank God for falling standards!
> -- Relieved mother, finding out her son's good grades (as reported in the *Daily Mail*, 2004)

Say what you want about the Ten Commandments, you must always come back to the pleasant fact that there are only ten of them.
> -- H.L. Mencken (1880-1956)

Factoid: The Bible doesn't actually specify how many wise men there were—just that they brought 3 gifts.

I just spent two weeks doing a jigsaw puzzle. I was quite pleased with myself because on the box it said "6-8 years".
> -- Les Blake

A man with one watch knows what time it is. A man with two watches is never sure.
> -- Albert Einstein (1879-1955)

Even a stopped clock tells the right time twice a day.
> -- Anonymous

You mean now?
> -- Yogi Berra (1925-2015) when asked for the time of day

Those who speak most of progress measure it by quantity and not by quality.
 -- George Santayana (1863-1952)

We need to make the pie higher.
 -- George W. Bush

They're called "countermeasures" because they are supposed to counter the bad guys, and because you are supposed to measure them.
 -- Anonymous

"Better make it six, I can't eat eight."
 -- Baseball player Dan Osinski, when a waitress asked if he wanted his pizza cut into 6 or 8 slices

You can only model what you can measure.
 -- Bob Conlin

News correction: Because of an editing error, a recipe last Wednesday for meatballs with an article about foods to serve during the Super Bowl misstated the amount of chipotle chillies (sic) to be used. It is one or two canned chillies (sic), not one or two cans.
 -- New York Times

If you suck on a tit the movie gets an R rating. If you hack the tit off with an axe it will be PG.
 -- Jack Nicholson

The shortest distance between any two points is under construction.
 -- Noelie Altito

There's a remote tribe that worships the number zero. Is nothing sacred?
 -- Les Dawson

There's no sense in being precise when you don't even know what you're talking about.
 -- John von Neumann (1903-1957)

This businessman found himself in a large city with a few hours to kill during one of his trips, so he decided to go downtown to visit the Natural History Museum. There in the main atrium was a huge dinosaur skeleton. After staring at this impressive creature for a while, he decided to strike up a conversation with a nearby security guard. "How old do you think that thing is?" the man asked. "Oh, that dinosaur is 70 million and 29 years old," said the guard. "70 million and 29!" exclaimed the man, "How do you know that?" "Well," said the guard, "when I first started working here 29 years ago, they told me it was 70 million years old."

2 + 2 = 5 for extremely large values of 2
    -- Bumper sticker

Security effectiveness is not measured by how much productive work gets impeded, nor by how much employees get inconvenienced.
    -- Anonymous

Actual courtroom testimony:
Q: Meaning no disrespect, sir, but you're 80 years old, wear glasses, and don't see as well as you used to. So tell me, just how far can you see?
A: Well, I can see the moon. How far is that?

Yogi Berra was sore after a game, so he went into the training room and jumped into the whirlpool. He immediately let out a yelp and jumped out. The trainer turned around and said, "What's the matter, Yogi? Is the water too hot?" "I don't know," said Yogi, "How hot is it supposed to be?"

The average human has about one breast and one testicle.
    -- From *Statistics 101*

Any discipline that involves the use of statistics is not cool by definition.
    -- Lalith Vipulananthan

Definition—Design Basis Threat (DBT): Maybe if we string together 3 disjoint adjectives, the blatantly obvious—that you ought to design your security with the adversary in mind—will seem profound.
    -- *Devil's Dictionary of Security Terms*

Design Basis Threat is not something you test against, it's a technique for allocating resources.
    -- Roger Johnston

We often demonstrate prototype cargo security systems by using a toy truck that is a 1:32 die-cast scale model. We learned early on that if we don't include something that looks like a truck, the people interested in cargo security get confused about what it is we are trying to demonstrate. On one occasion, a guy we were briefing

said that he liked the concept, but that our security hardware was simply too large. We were initially confused at this statement since our prototype was only a couple of cubic inches in volume, and could be further miniaturized. We eventually figured out his problem. It took a long time to explain to him—and I'm not sure he ever fully grasped the concept—that our prototype was at a 1:1 scale, while the truck was 1:32, i.e., a real truck is much bigger but our hardware would not be.
-- Roger Johnston

We have some great activities in our village games, like the 100 meter hurdles, which are a bit stupid because none of us can jump that high.
-- Bennett Arron

I don't know. I'm not in shape yet.
-- Yogi Berra (1925-2015), when asked his cap size

I've been big ever since I was little.
-- William "The Refrigerator" Perry

Q: What did you get on your SAT test?
A: Nail polish.
-- Interviewer and response from Jennifer Lopez

Next week I have to take my college aptitude test. In my high school, they didn't even teach aptitude.
-- Tony Banta, *Taxi*

A can of Campbell's Pork and Beans, in which the "Pork" ingredient has always been infinitesimal, has apparently accomplished the impossible, unless they're just using smaller beans: The label says, "Now with more beans!"

Some unusual units of measure:
barn: A unit of area equal to 1E-28 square meters (10 raised to the -28th power) used by scientists in measuring the interaction cross section of atomic nuclei. Called this because this area is "as big as a barn" on the nuclear scale. An "outhouse" is defined as 1E-34 square meters and a "shed" is 1E-52 square meters.
beard-second: The average length a man's beard grows in 1 second, about 10 nanometers.
Helen: Helen of Troy's beauty supposedly launched a thousand ships. The degree of beauty needed to launch one ship is thus a milliHelen. According to this unit's

developer, David Lance Goines, a picoHelen is the unit of beauty that inspires men to "barbecue a couple of steaks".

Wheaton: The actor Will Wheaton was one of the first actors to have a huge Twitter following. When he hit 500,000 followers (he eventually reached over 1.75 million) that number was called a "Wheaton".

Warhol: Being famous for 15 minutes. Andy Warhol once said that, "In the future, everyone will be famous for fifteen minutes." A kiloWarhol is being famous for about 10 days.

Scoville: A unit of measure for chile hotness named after its inventor, Wilbur Scoville. A habanero has a Scoville rating of 200,000, meaning that it has to be diluted 1:200,000 before the heat is not detectable.

smidgen: Officially, 1/32 of a teaspoon.

Definition—security standard: A committee of special interests tries to legitimize poor practice and sloppy terminology through formal means.
-- *Devil's Dictionary of Security Terms*

The great thing about standards is that there are so many to choose from.
-- old engineering joke

It's Standard Maxim: As a general rule of thumb, about two-thirds of security standards or certifications make security substantially worse.

30 - Safety

Key Points

• **As with security, there is safety compliance and then there is safety. The two only partially overlap.**

• **Having a good Safety Culture matters as much as having a good Security Culture.**

Accidents don't just happen. They must be carelessly planned.
> -- Anonymous

Safety never takes a holiday.
> -- From the movie, P*aul Blart: Mall Cop* (2009)

I childproofed my house but the kids still get in somehow.
> -- Anonymous

In case of contact [with this chemical], immediately wash skin with soap and copious amounts of water. If swallowed, wash out mouth with water provided the person is conscious, and call a physician.
> -- Material Safety Data Sheet (MSDS) for sucrose (table sugar)

In the same way that the Iranian theocracy has raised the most irreligious generation that the mullahs have ever seen, the Safety Mullahs have bred indifference to all but the most strident warning labels.
> -- Derek Lowe

Printing on a dashboard sunshield: "Warning: Do not drive with sunshield in place."

When we think of the major threats to our national security, the first to come to mind are nuclear proliferation, rogue states and global terrorism. But another kind of threat lurks beyond our shores, one from nature, not humans—an avian flu pandemic.
> -- Barak Obama

Go the place of danger, for there you will find safety.
-- Chinese proverb

Out of this nettle, danger, we pluck this flower, safety.
-- William Shakespeare (1564-1616), *King Henry IV*. Part I. Act 2, Scene 3.

Ride in our taxis, and you'll never walk again!
-- Advertising slogan for an Irish taxicab service

Why did Kamikaze pilots wear helmets?
-- Anonymous

Why do they use sterilized needles for death by injection?
-- Anonymous

Blink your eyelids periodically to lubricate your eyes.
-- Page 16 of the Hewlett-Packard Environmental, Health, & Safety Handbook for Employees

If your eye hurts after you drink coffee, you might want to take the spoon out of the cup.
-- Norm Crosby

The effectiveness of safety rules should not be measured by the extent to which productive work is impeded, but by the reduction in the ratio of harm to hours worked.
-- Anonymous

On a Korean Kitchen Knife: "Warning: Keep out of children."

Ah, this is obviously some strange usage of the word 'safe' that I wasn't previously aware of.
-- Arthur Dent in *The Hitchhikers Guide to the Galaxy*, Douglas Adams (1952-2001)

Mr. Burns: The watchdog of public safety. Is there any lower form of life?
-- The Simpsons (1989)

The bridge of the Starship Enterprise is always exploding with high voltage discharges when under attack. What in the world were the designers thinking by

running high voltage through the control consoles? What's wrong with 5-volt TTL logic or even photonics?
-- Roger Johnston

Warning on shin guards for bicyclists: "Shin pads cannot protect any part of the body they do not cover!"

Warning label on a CD player: "Do not use the UltraDisc 2000 as a projectile in a catapult."

Actual consumer warning label on a tractor: DANGER: AVOID DEATH.

On a laser printer cartridge: "Warning. Do not eat toner."

If the safety pin were invented today, it would have 2 transistors, a regulator, an off-and-on switch, and require a service check every 6 months.
-- *Bits & Pieces*

"...while consciousness is obviously necessary...it is not itself a job function."
-- EEOC argument in court that Conrail should not deny a train dispatcher's job to an employee who continuously blacks out

I have every sympathy with the American who was so horrified by what he had read about the effects of smoking that he gave up reading.
-- Henry G. Srauss

Factoid: In the history of aviation, a total of 13 people are known to have fallen out of airplanes at substantial altitude without any means for deceleration and yet survived after falling to Earth.

31 - Stupidity/Ignorance

Some Key Points:

• **It's everywhere, and it's dangerous.**

I'm sure the universe is full of intelligent life. It's just been too intelligent to come here.
    -- Arthur C. Clarke (1917-2008)

Nothing in all the world is more dangerous than sincere ignorance and conscientious stupidity.
    -- Martin Luther King, Jr. (1929-1968)

Stupidity has a knack of getting its own way.
    -- Albert Camus (1913-1960)

Only two things are infinite, the universe and human stupidity, and I'm not sure about the former.
    -- Albert Einstein (1879-1955)

What's a Walmart? Do they sell, like, wall stuff?
    -- Paris Hilton

Sundance is weird. The movies are weird. You actually have to think about them when you watch them.
    -- Pop singer Britney Spears

I am patient with stupidity but not with those who are proud of it.
    -- Edith Sitwell (1887-1964)

Strange as it seems, no amount of learning can cure stupidity, and higher education positively fortifies it.
    -- Stephen Vizinczey

I pity the fool.
    -- Mr. T

Why does mineral water that has "trickled through mountains for centuries" go out of date next year?
    -- Anonymous

There is nothing worse than aggressive stupidity.
    -- Johann Wolfgang von Goethe (1749-1832)

A stupid man's report of what a clever man says is never accurate because he unconsciously translates what he hears into something he can understand.
    -- Bertrand Russell (1872-1970)

A genius in a room full of idiots is the idiot.
    -- Anonymous

I've never really wanted to go to Japan simply because I don't really like eating fish, and I know that's very popular out there in Africa.
    -- Pop singer Britney Spears

God, deliver me from a man of one book.
    -- English proverb

I love California, I practically grew up in Phoenix.
    -- Dan Quayle

I get to go to lots of overseas places, like Canada.
    -- Pop singer Britney Spears

The hard part about being a bartender is figuring out who is drunk and who is just stupid.
    -- Richard Braunstein

If you're listening to a rock star in order to get your information on who to vote for, you're a bigger moron than they are.
    -- Alice Cooper

All morons hate it when you call them a moron.
    -- J.D. Sallinger (1919-2010)

I am amazed at radio DJ's today. I am firmly convinced that AM on my radio stands for Absolute Moron. I will not begin to tell you what FM stands for.
    -- Jasper Carrott

All the stuff I was taught about evolution and embryology and Big Bang Theory, all that is lies straight from the pit of hell.

-- Georgia Congressman Paul Broun, a physician and member of the House Committee on Science, Space, and Technology

32 - Uh…How's That Again?

Some Key Points:

• **There can be wisdom in dumbness. Or not.**

Radisson Welcomes
Emerging Infectious Diseases
    -- Sign outside a Radisson Hotel

Actual courtroom testimony:
Q: Do you recall the exact distance?
A: That he was from me? Or I was from him?

I don't think anybody should write his autobiography until after he's dead.
    -- Samuel Goldwyn (1879-1974)

Cleveland Indians catcher Harry Chiti was once traded for himself. He was traded for a player to be named later, and (2 months later) he turned out to be that player.

As we know, there are known knowns. There are things we know we know. We also know there are known unknowns. That is to say, we know there are some things we do not know. But there are also unknown unknowns, the ones we don't know.
    -- Donald Rumsfeld

Wow…if only a face could talk!
    -- Sportscaster John Madden during the Super Bowl

And for those of you watching who haven't television sets, live commentary is on Radio 2.
    -- Television commentator David Coleman

Actual courtroom testimony:
Attorney: Are you sexually active?
Witness: No, I just lie there.

Uh...How's That Again?

Please feel free to contact me with other matters that are of importance to you. I am honored to serve as your representative in the U.S. Congress. I think you're an asshole.
Sincerely,
JoAnn Emerson
Member of Congress
-- Actual letter sent by Rep. JoAnn Emerson who denied knowing where the last sentence came from. She pledged to launch an immediate investigation.

The ignorant mind is too often a fresh face. The ignorant mind is too often—mine—young mind is too often something you can lose.
-- George H.W. Bush

If the Phone Doesn't Ring, It's Me.
-- Song title by Jimmy Buffett

Write music like Wagner, only louder.
-- Directions from movie mogul Samuel Goldwyn (1879-1974)

Wagner's music is better than it sounds.
-- Mark Twain (1835-1910)

For the third goal, I blame the ball.
-- Saudi goalkeeper Mohammed Al-Deayea

You guys pair up in groups of three, then line up on a circle.
-- Instructions given to his football players by legendary Florida State football coach Bill Peterson

I'm not a rocket surgeon!
– Tattoo artist and reality TV star Kat Von D

If Stanford is a number 12 seed, then I'm a left-handed ham sandwich.
-- Wimp Sanderson

Newspaper correction: An article about Lord Lambton ("Lord Louche, Sex King of Chiantishire," News Review, January 7) falsely stated that his son Ned (now Lord Durham) and daughter Catherine held a party at Lord Lambton's villa, Cetinale, in 1997, which degenerated into such an orgy that Lord Lambton banned them from Cetinale for years. In fact, Lord Durham does not have a sister called Catherine (that is the name of his former wife), there has not been any orgiastic party of any kind and Lord Lambton did not ban him (or Catherine) from Cetinale at all. We apologise sincerely to Lord Durham for the hurt and embarrassment caused.

It's not true that I like human flesh. It's much too salty for me.
    -- Idi Amin Dada (1925-2003)

Walmart's new line of makeup for preteens, called geoGirl, includes anti-aging ingredients.

A complete ripoff of *Jaws*.
   -- One star review on Amazon.com of Herman Melville's *Moby Dick*

Actual courtroom testimony:
Witness (a Physician): He was probably going to lose the leg, but at least maybe we could get lucky and save the toes.

It's a sociopolitical statement about the state of our materialistic, hype-driven society, you know?
    -- Actress Rachael Leigh Cook discussing her movie, *Josie and the Pussycats*, based on a Saturday morning kid's TV cartoon show

33 - Crime & Violence

Some Key Points:

• **It's ubiquitous.**

Chicago Tribune Headline: "Headless Blonde Found in Thames"

Outside of the killings, Washington has one of the lowest crime rates in the country.
-- Marion Barry, Washington D.C. Mayor

If crime went down 100%, it would still be 50 times higher than it should be.
-- Councilman John Bowman commenting on the high crime rate in Washington

You can get much farther with a kind word and a gun than you can with a kind word alone.
-- Attributed to Al Capone (1899-1947)

On an actual insurance claim: The Insured was fishing at Cold Springs Harbor and hooked a large fish. The fish, however, swam off with her rod and reel. The claim is for theft.

Factoid: It is illegal in Indiana to open a can of food using a gun.

Factoid: It is illegal in Virginia to call somebody up on the phone and then say nothing.

When asked by a reporter why he robbed banks, bank robber William "Willie" Sutton (1901-1980) supposedly said, "because that's where the money is." [Sutton eventually claimed the reporter made up the quote.]

Holdup note handed to a bank security guard by wannabe band robber Ronnie Darnell Bell: "This is a bank robbery of the Dallas Federal Reserve Bank of Dallas. Give me all the money. Thank you. Ronnie Darnell Bell."

It is better to risk saving a guilty person than to condemn an innocent one.
-- Voltaire (1694-1778)

A thief is sorry that he is going to be hanged, not that he is a thief.
-- English proverb

We apologize for the error in last week's paper in which we stated that Mr. Arnold Dogbody was a defective in the police force. We meant, of course, that Mr. Dogbody is a detective in the police farce.
-- Correction notice in the British newspaper, *Ely Standard*

You'll meet someone. Someone very special. Someone who won't press charges.
-- From the movie, *Addams Family Values* (1993)

I'd kill for a Nobel Peace Prize.
-- Steven Wright

Two young women were in a car, speeding down the highway at well over the speed limit. "Hey," asked the brunette behind the wheel, "are any cops following us?" The blonde passenger turned around for a look. "Yeah, I see a police car following us now!" "Oh, No!" moaned the brunette. "Are his flashers on?" The blonde turned around again to look. "Yep…nope…yep…nope…yep…"

In a closed society where everybody's guilty, the only crime is getting caught. In a world of thieves, the only final sin is stupidity.
-- Hunter S. Thompson (1939-2005)

Inspector Jacques Clouseau: The good cop/bad cop routine is working perfectly. Ponton: You know, usually two different cops do that.
-- From the movie, The Pink Panther (2006)

I heard some good news today. The FBI and the CIA are going to start cooperating…And if you don't know the difference between them, the FBI bungles domestic crime, while the CIA bungles foreign crime.
-- David Letterman

The high school called the other day and said they caught my kid cheating on his chemistry exam. I reassured them it was ok because he wants to be an FBI forensics scientist when he grows up.
-- Anonymous

It's kind of pathetic that the last time the FBI was a competent crime fighting organization, it was led by a paranoid, blackmailing, transvestite thug who hated blacks, Jews, and the U.S. Constitution.

-- Anonymous

Always remember to pillage BEFORE you burn.
    -- Anonymous

Advice from a crime expert on television:
"Don't go into darkened parking lots unless they are well lighted."

Now they show you how detergents take out bloodstains, a pretty violent image
there. I think if you've got a T-shirt with a bloodstain all over it, maybe laundry isn't
your biggest problem.
    -- Jerry Seinfeld.

When you go into court, you are putting yourself in the hands of 12 people that
weren't smart enough to get out of jury duty.
    -- Anonymous

If crime fighters fight crime and firefighters fight fires, what do freedom fighters
fight?
    -- George Carlin (1937-2008)

I'm the kind of crazy you weren't warned about because no one knew this level
existed.
    -- Anonymous

I went to a fight the other night and a hockey game broke out.
    -- Rodney Dangerfield (1921-1997)

Actual courtroom testimony:
Q: Have you lived in this town all your life?
A: Not yet.

Whoever said that money can't buy friends obviously never brought donuts to the
office.
    -- Wendy Weiner Runge

The key is to commit crimes so confusing that police feel too stupid to even write a
crime report about them.
    -- Randy K. Milholland

Actual police incident:
Police Officer to purse snatching suspect in the lineup: Put your baseball cap on the other way, with the bill facing front.
Suspect: No, I'm gonna put it on backwards. That's the way I had it on when I took the purse.

If you choke a smurf, what color does it turn?
-- Steven Wright

There is a great streak of violence in every human being. If it is not channeled and understood, it will break out in war or in madness.
-- Sam Peckinpah

The police officer reported to his watch commander about the difficulties he was having interrogating a witness. "Did you browbeat him, yell at him, intimidate him, and ask every pointed question you could come up with?" asked the watch commander. "I sure did," said the officer, "but all that happened is that he mumbled, 'Yes, dear, you're right.' and then dozed off!"

Things aren't right. If a burglar breaks into your home and you shoot him, he can sue you. For what, restraint of trade?
-- Bill Maher

Gentlemen, you can't fight in here! This is the War Room.
-- From the movie, *Dr. Strangelove or How I Learned to Stop Worrying and Love the Bomb* (1964)

After the meek inherit the Earth, I think we should just kick their butts and take it from them.
-- Jim Rosenburg

I can picture in my mind a world without war, a world without hate. And I can picture us attacking that world, because they'd never expect it.
-- Jack Handey

I'm not in favor of self-sacrifice. Or any human sacrifice for that matter. Too bloody.
-- Anonymous

If it weren't for pickpockets, I'd have no sex life at all.
-- Rodney Dangerfield (1921-2004)

Did you give consent for someone to steal or damage your property?
-- Crime victim report form, Madison (Wisconsin) Police Department

My husband gave me a necklace. It's fake. I requested fake. Maybe I'm paranoid, but in this day and age, I don't want something around my neck that's worth more than my head.
-- Rita Rudner

Harry Solomon (reading the paper): Here's a job that I can do: "Police are seeking third gunman." Tomorrow, I'm gonna march over to the police station and show them that I'm the man they're looking for.
-- *Third Rock from the Sun*

I haven't committed a crime. What I did was fail to comply with the law.
-- David Dinkins, Mayor of New York City

I was provided with additional input that was radically different from the truth. I assisted in furthering that version
-- Colonel Oliver North, from his Iran-Contra testimony

I deny the allegations and I defy the alligators!
-- Indicted Chicago Alderman

The day is for honest men, the night for thieves.
-- Euripides (484 – 406 BC)

A crime which is the crime of many, none avenge.
-- Lucan (39 - 65 AD)

Crime is naught but misdirected energy.
-- Emma Goldman (1869-1940)

Nothing is more destructive of respect for the government and the law of the land than passing laws which cannot be enforced. It is an open secret that the dangerous increase of crime in this country is closely related with this.
-- Albert Einstein (1879-1955)

Organized crime in America takes in over forty billion dollars a year and spends very little on office supplies.
-- Woody Allen

Behind every great fortune there is a crime.
  -- Honore de Balzac (1799-1850)

You don't want another Enron? Here's your law: If a company, can't explain in ONE SENTENCE what it does, IT'S ILLEGAL!
  -- Louis Black

Indeed, history is nothing more than a tableau of crimes and misfortunes.
  -- Voltaire (1694-1778)

History is just one damn thing after another.
  -- Attributed to many people

No Trespassing signs make no sense. Trespassing is access to private property without right or permission. The owner can't grant the right to trespass; if trespassing is allowed, it isn't trespassing!

Many police practices may be useful for fighting crime--preventive detention or coerced confessions, for example--but because they are unconstitutional they cannot be used, no matter how effective.
  -- Federal Judge Sjhira Scheindlin in her 2013 ruling on New York City's 'stop and frisk' policy. (Actually, stop and frisk never was very effective.)

My mother used to say that there are no strangers, only friends you haven't met yet. She's now in a maximum security twilight home in Australia.
  -- Dame Edna Everage

I think the inventor of the piñata may have had some unresolved donkey issues.
  -- Dan Johnson

Illegal aliens have always been a problem in the United States. Ask any Indian.
  -- Robert Orben

Actual courtroom testimony:
Q: James shot Tommy Lee?
A: Yes.
Q: Then Tommy Lee pulled out his gun and shot James in the fracas?
A: No, sir, just above it.

Always go to other people's funerals. Otherwise they might not come to yours.

-- Yogi Berra (1925-2015)

Actual courtroom testimony:
Q: When was the last time you saw the deceased?
A: At his funeral.
Q: Did he make any comments to you at that time?

They say you shouldn't say nothing about the dead unless you can say something good. He's dead. Good.
    -- Moms Mabley (1894-1975)

It may be true that the law cannot make a man love me, but it can stop him from lynching me, and I think that's pretty important.
    -- Martin Luther King, Jr. (1929-1968)

To me, boxing is like a ballet, except there's no music, no choreography, and the dancers hit each other.
    -- Jack Handey

There's so much comedy on television. Does that cause comedy in the streets?
    -- Dick Cavett

Factoid: According to the Bureau of Justice Statistics, 1.7 million Americans are the victims of violent crime in the workplace each year, and 800 are killed.

The reason there is so little crime in Germany is that it's against the law.
    -- Alex Levin

For every fatal shooting, there are roughly three non-fatal shootings. And, folks, this is unacceptable in America.
    -- George W. Bush

In all fairness, there were others who didn't get along with him. He's a very difficult man to work for.
    -- FBI Agent Anthony Nelson discussing a case where an employee put cyanide into his boss's water cooler

Finally I've found something that combines my love of helping people with my love of hurting people.
    -- Homer Simpson, on his new job as a vigilante cop

English television talk show with guests Paul Merton and Member of Parliament Glenda Jackson:
Host: Do you remember your school motto?
Glenda Jackson: [Unsure of whether the host is asking her or Paul Merton] Who are you looking at?
Paul Merton: That must have been a tough school!

Actual courtroom testimony:
Q: Were you freebasing the cocaine?
A: No. I bought it.

The last thing you want is for somebody to commit suicide before executing them.
    -- Gary Deland, Utah Director of Corrections, commenting on a special
    holding cell for death row prisoners awaiting execution by firing squad

Actual 911 recorded call:
911: 911. What's your emergency?
Caller: Someone broke into my house and took a bite out of my ham-and-cheese sandwich.
911: Excuse me?
Caller: I made a ham-and-cheese sandwich and left it on the kitchen table, and when I came back from the bathroom, someone had taken a bite out of it.
911: Was anything else taken?
Caller: No. But this has happened to me before, you know, and I'm sick and tired of it.

See, that is one of my pet peeves. You should be able to go and pick out one fan a game, and just beat the hell out of him.
    -- Basketball player Charles Barkley

Toronto Sun television program guide: Perry Como's Christmas Special--The members of a Greek family are murdered systematically in a bizarre fashion.

If you see a Bulgarian on the street, beat him. He will know why.
    -- Russian adage

Factoid: Traditionally, the most frequently shoplifted book is the Bible.

The best way to get a bad law repealed is to enforce it strictly.
    -- Abraham Lincoln (1809-1865)

A liberal is a conservative who's been arrested. A conservative is a liberal who's been mugged.
   -- Wendy Kaminer

When the biggest, richest, glassiest buildings in town are the banks, you know that town's in trouble.
   -- Edward Abbey (1927-1989)

34 - Dogs, and What They Can Teach Us

Some Key Points:

- **Woof!**

- **Dogs can teach us about perception, cognition, and security stupidity.**

Your goal in life ought to be to become the person your dog already thinks you are.
-- Anonymous

Outside of a dog, a book is man's best friend. Inside of a dog, it's too dark to read.
-- Groucho Marx (1890-1977)

We can learn a lot from dogs. They enthusiastically embrace life, live in the moment, work hard and play hard, are full of love and loyalty, don't hold grudges, and never worry about the future or second guessing themselves.
-- Anonymous

Dogs live in the moment. They don't regret the past or worry about the future. If we can learn to appreciate and focus on what's happening in the here and now, we'll experience a richness of living that other members of the animal kingdom enjoy.
-- Cesar Millan

A door is what a dog is perpetually on the wrong side of.
-- Ogden Nash (1902-1971)

Dogs bark at what they don't understand.
-- Heraclitus (535 - 475 BC)

Barking dogs don't bite.
-- Chinese proverb

It's not the size of the dog in the fight, it's the size of the fight in the dog.
-- Mark Twain (1835-1910)

A man was very proud of his guard dog and would let the dog roam free in his backyard to show the world his house was guarded. One day a woman knocked at his door. "Is that your big dog outside?" Wondering how she had got past him he said, "Yes why?" She said, "I'm sorry but my dog just killed him!" "What??" roared the man, "What kind of dog have you got?" "A Pekingese," replied the woman. "A Pekingese? How could that little thing kill my monster guard dog?" "I think it got stuck in his throat," replied the woman.

Try our new horse-flavored dog food! Your dog will love it!
    -- TV commercial spoof

My dog is half pit bull and half Poodle. It isn't much good as a guard dog, but it sure is a vicious gossip.
    -- Craig Shoemaker

My only serious brush with the law occurred because I was out of town on travel one summer. Unbeknownst to me, my girlfriend had moved her dog into my backyard because it was cooler and the dog was suffering from the heat spell. (I lived 1000 feet higher in elevation.) Well, the dog was unhappy about being left alone in a new place and barked all night long. I ended up getting a citation for having a nuisance dog. When my day in court arrived, I explained to the judge that this all happened without my knowledge, that it was a one-time event, and that, in any case, the dog had unfortunately died from a heart attack (apparently unrelated to the heat). The judge ruled that I was guilty, but that if the dog failed to create any additional disturbances over the course of the next 90 days, the charges would be dropped. Perhaps not surprisingly, it turned out that the dead dog engaged in no annoying barking during those 90 days, and I was off the hook.
    -- Roger Johnston

Art Linkletter used to host a popular television show called "Kids Say the Darndest Things". He would interview children to get their unique and often funny perspective on the world. One day before the show, the kids who were to be interviewed were running around and having fun on the set. All except for one little boy who was by himself over in the corner, looking sad. So Art went over to talk to him. "You look unhappy," said Art. "Is everything ok?" "Well," said the little boy, "My dog died last week." "Gee, I'm sorry to hear that," said Art. "But don't worry. I'm sure someday you will see your dog in heaven." The kid looked up at Art, thought for a moment, then said, "Well....ok. But what does God want with a dead dog?"

My husband is a lot like my dog. Both snarf down their dinner, and both are afraid of the vacuum cleaner.

-- Anonymous

Factoid about the (for real) ancient sport of kadabbi: There are 7 players on each team. One member of the team is sent into the other team's half of the court. His job is to tag as many opposing players as possible while holding his breath and continuously chanting "badabbi" to prove he isn't inhaling. The game was demonstrated at the 1936 Berlin Olympics.

Home computers are being called upon to perform many new functions, including the consumption of homework formerly eaten by the dog.
    -- Doug Larson

Dogs are the leaders of the planet. If you see two life forms, one of them's making a poop, the other one's carrying it for him, who would you assume is in charge?
    -- Jerry Seinfeld

I loathe people who keep dogs. They are cowards who haven't got the guts to bite people themselves.
    -- August Strindberg (1849-1912)

The Saint Bernards work best in teams of at least three dogs. They are sent out on patrols following storms, and they wander the paths looking for stranded travelers. If they come upon a victim, two dogs lie down beside the person to keep him warm; one of the two licks his face to stimulate him back to consciousness. Meanwhile, another dog will have already started back to the hospice to sound the alarm.
    -- Stanley Coren

Diplomacy is the art of saying 'nice doggie' until you can find a rock.
    -- Will Rogers (1879-1935)

When a dog runs at you, whistle for him.
    -- Henry David Thoreau (1817-1862)

Factoid: Only 3 animals walk by moving both their left feet, then their right feet: giraffes, camels, and cats. This motion allows for good speed and agility, and lets them move quietly.

I used to have 2 mutts, one a fairly large Lab-mix and the other a small, white American Eskimo (Spitz) named Zara. Now Zara was, by all accounts, one of the dumbest dogs on the planet. In his later years, he developed dog cognitive disorder, but we couldn't see that it made a whole lot of difference.

Dogs

We kept the dogs in the backyard. One day, I looked out the window and saw a large black bear in the backyard. A drought was ongoing and the bear was probably hungry. Seeing the bear, the Lab-mix, intelligently enough, was desperately clawing at the back door to be allowed inside, safe and away from the bear. Zara, on the other hand, took one look at the bear from about 80 feet away, put his head down, and silently charged it. It all happened so fast that I couldn't call out, but I instantly figured that Zara was one dead dog. A flick of the wrist by the bear would probably kill him.

Now the bear was likely familiar with being hassled by dogs, but this situation somehow just didn't compute: a white blur silently charging him made no sense. I saw this almost human look of terror on the bear's face. Before Zara could reach him, the bear spun around, leapt over the 4 foot chain link fence, careened off the roof of our neighbor's metal storage shed, and sprinted off. (Bears can run remarkably fast when they want to.)

The lesson I took from this incident is that you can accomplish great deeds if only you are too stupid to recognize your own limitations.

-- Roger Johnston

This same dog taught me about how little we understand pattern recognition and cognition. Zara hated people who wore hats. He didn't even like it when I wore a hat. Perhaps he had been frightened or hassled by somebody in the past who wore a hat.

Anyway, despite being approximately the world's dumbest dog, he could determine within a fraction of a second whether somebody was wearing a hat. It didn't take a lot of mental computation time. This was being done with neurons that operate (even in a normal dog or a person) at remarkably slow processing speeds. He could instantly recognize hats without any context or understanding of hats or what they are used for. (He didn't, after all, wear hats.)

Zara never got confused by the person being tall or having big hair, and he could do his hat recognition trick without the benefit of much color perception to help differentiate hat from hair or skin. In fact, canine vision systems are quite poor compared to human eyesight. Dog vision is really more like a motion sensor than vision as we experience it.

Despite Zara's stupidity and various limitations, few (if any) computer programs could match his ability to spot hats so quickly and reliably, and in novel situations and varying illumination. There is clearly a lot going on inside a brain—even a dumb dog's brain—than we are anywhere close to understanding. This is one of the reasons that blink comparators work so well for comparing "before" and "after" images; the blink comparison relies on the remarkable pattern recognition and cognition of the human brain.

-- Roger Johnston

As a vulnerability assessor, I see a lot of dumbness in security. There seems to be 3 main kinds. There is the dumbness associated with inadvertent errors—forgetting to lock the door or to install the latest virus definitions or software updates, for example. These inadvertent blunders can have serious security implications, but I'm not sure I have any profound countermeasures to offer.

Then there are those people and organizations who are simply too stupid to ever have good security. They are beyond all hope or help, but I truly believe they are very much in the minority.

The largest category of security dumbness seems to be those people and organizations that, at least in theory, should have sufficient common sense and intelligence to have good security. They just don't choose to apply it to security. They fit into what I call the "Dog Snot Model of Security".

The Dog Snot Model of Security—as in "dumber than dog snot"—arose from my observations of my two dogs, Mr. Rigley and Carly. They are rescue dogs. Mr. Rigley got in trouble with the Houston police, and Carly was rescued from solitary confinement in Santa Fe after a parvo outbreak. They are both probably (short hair) Dutch Shepherds, a breed not formally recognized in the United States but popular in the Netherlands. Dutch Shepherds often have a brindle color. They were bred by Dutch farmers to herd sheep and pull little carts in the fields.

As is common for the breed, our 2 dogs are very smart. Carly, for example, understands about 200 words. Dutch Shepherds are sometimes used as police and military dogs because of their intelligence and modest size (50-55 pounds).

Dogs

Like most working dog breeds, if you don't give them enough to do, they will find a job. Mr. Rigley and Carly have decided that their main role in life is to guard the backyard from the appearance of Max the barn cat. In the dogs' view, this is a kind of security function.

We have several floor-to-ceiling picture windows in the living room looking out over the backyard. When the dogs see Max, typically cutting through the backyard on his way to the barn, the dogs go berserk, and throw themselves at the windows, barking and clawing like crazy. This is all rather stupid because they have access to an automated dog door that would let them out to bark at Max up close, and maybe even catch him depending on his route. Instead, they foolishly and futilely slam against the windows—much like otherwise sensible people or organizations who behave stupidly and in counterproductive ways when it comes to security.

One might argue that perhaps the dogs simply forget the windows are there, being transparent. But this clearly isn't the case because enough dried dog snot ends up on the windows that they are often not very transparent. Thus the name of the model.

-- Roger Johnston

35 - Physical Security/Building & Facility Security

Some Key Points:

- **Physical Security: not really a field at all?**

I would've thought that physical security was a solved problem.
      -- Jon Warner

Physical security is not so much a discipline as it is a collection of special interests.
      -- Roy Lindley

This house is wide open and you people have no clue what real security is or what it takes to achieve it.
      -- From the movie, *The Bodyguard* (1992)

In a real field, you can get a degree from a major research university in the topic. There are fundamental principles, theories, models, metrics, meaningful standards, and intelligent guidelines. Rigor, critical thinking, and creativity abound. There are frequent R&D conferences and numerous peer-reviewed journals. Physical security falls short on all of these things.
      -- Roger Johnston

Factoid: If you do major computer hacking and get caught, you may go to jail. After you get out, you can become a well-paid cyber security consultant. With physical security, in contrast, once you are charged with a crime, you most likely will never be allowed to work in the field or use what you have learned to improve security.

In cyber security, vulnerabilities are assumed to be prevalent and never fully avoidable. With physical security, in contrast, vulnerabilities are usually taken to mean that somebody has been screwing up. Thus, security managers and organizations are hesitant to look for, acknowledge, or fix physical security vulnerabilities.
      -- Roger Johnston

You've done a nice job decorating the White House.

-- Pop star Jessica Simpson, during a VIP tour of the White House upon being introduced to the U.S. Secretary of the Interior

I might be in the basement. I'll go upstairs and check.
   -- M.C. Escher (1898-1972)

Housework, if you do it right, will kill you.
   -- Erma Bombeck (1927-1996)

One of the laws of paleontology is that an animal which must protect itself with thick armor is degenerate. It is usually a sign that the species is on the road to extinction.
   -- John Steinbeck (1902-1968)

36 - Inventory is not Security!

Some Key Points:

• **Confusing Inventory with Security: An almost universal mistake.**

• **This is fundamentally why the Global Positioning System (GPS), Radio Frequency Identification Devices (RFIDs), and Contact Memory Buttons (CMBs) are easy to spoof and make little sense for many security applications.**

Definition—Inventory: Counting and locating our assets. Innocent misplacement errors by insiders may be detected, but we can conclude little about theft. Detecting theft is the job of security—a completely different function that consciously seeks to counter the bad guys.

If an inventory system makes no significant effort to deal with deliberate spoofing, it's not a security system. And you can draw no meaningful conclusions from it about theft.
    -- Roger Johnston

Not everything that can be counted counts, and not everything that counts can be counted.
    -- Attributed to Albert Einstein (1879-1955)

Mission Creep Maxim: Any given device, system, or program that is designed for inventory will very quickly come to be viewed—quite incorrectly—as a security device, system, or program.

In some ways, RFIDs (Radio Frequency Identification Devices) are less secure than even printed paper barcodes. For example, it's easy to spoof an RFID reader from a distance using hobbyist kits you can get from the Internet, but it's challenging to do that to an optical bar code reader.
    -- Roger Johnston

It's basically a bar code that barks.
    -- Robin Koh, speaking about RFIDs

There is a huge danger to customers using this (RFID) technology, if they don't think about security.

    -- Lukas Grunwald (creator of RFDump)

GPS installed in women's lingerie: See
http://www.dailymail.co.uk/news/worldnews/article-1082707/Outrage-chastity-belt-lingerie-fitted-GPS-tracking-system.html

37 - Designing & Choosing Security Products

Some Key Points:

• **Watch out for snake oil and for one-size-fits-all!**

• **Effective security cannot usually be retrofitted in.**

• **Don't ever make security plans based on the assumption that you will get optimal implementation of any security device, system, strategy, or program.**

• **Technology is a tool, not a solution, and you must use it wisely.**

With a lot of security devices, we have to wonder. Did anybody spend 60 seconds thinking about security?
    -- Roger Johnston

Factoid: The term "snake oil" refers to something that is fake, shoddy, or severely over-hyped. In the ancient world, medicines made from snakes were believed to have curative powers. Then, in 1880, John Greer began selling a snake oil "cure-all". The term, however, become widely used as a result of the 1893 Columbian Exhibition in Chicago. Clark Stanley ("The Rattlesnake King") sold his Snake Oil Liniment there. It was analyzed some years later and turned out to contain mineral oil, camphor, turpentine, beef fat, and chile powder—but no snake.

<u>Bob Knows a Guy Maxim</u>: Most security products and services will be chosen by the end-user based on purchase price plus hype, rumor, innuendo, hearsay, and gossip.

Extraordinary claims require extraordinary evidence.
    -- Carl Sagan (1934-1996)

<u>Contrived Duelism/Dualism Maxim</u>: The promoters of a security product meant to deal with any sufficiently challenging security problem will invoke a logical fallacy (called "Contrived Dualism") where only 2 alternatives are presented and we are pressured into making a choice, even though there are actually other possibilities. For example: "We found a convicted felon, gave him a crowbar, and he couldn't make the lock open after whaling on it for 10 minutes. Therefore, the lock is

secure." Another example, "Nobody in the company that manufacturers this product can figure out how to defeat it, and I bet you, Mr./Ms. Potential Customer [never having seen this product before in your life] can't think up a viable attack on the spot. Therefore, this product is secure."

Gold's Law: If the shoe fits, it's ugly.

Glazer's Law: If it says, "one size fits all", then it doesn't fit anybody.

Never purchase beauty products in a hardware store.
-- Miss Piggy

"Individual results may vary."
-- Weasel advertising cop out

We'll Worry About it Later Maxim: Effective security is difficult enough when you design it in from first principles. It almost never works to retrofit it in, or to slap security on at the last minute, especially onto inventory technology.

Invention is the mother of necessity.
-- Thorstein Veblen (1857-1929)

Schneier's First Maxim: The more excited people are about a given security technology, the less they understand (1) that technology and (2) their own security problems.

I bought an audio cleaning tape. I'm a big fan of theirs.
-- Kevin Gildea

Definition—video surveillance: A video system with such poor resolution, you couldn't recognize your own mother.

High-Tech Maxim: The amount of careful thinking that has gone into a given security device, system, or program is inversely proportional to the amount of high-technology it uses.

Shoot a few scenes out of focus. I want to win the foreign film award.
-- Billy Wilder (1906-2002)

This floodlight is capable of illuminating large areas, even in the dark.
-- Label on Komatsu floodlights

Don't go into darkened parking lots unless they are well-lighted.
  -- Advice from a crime specialist interviewed on television

Printing on a bottle of bathtub cleaner: For best results, start with clean bathtub before use.

Why pay a dollar for a bookmark? Why not use the dollar for a bookmark?
  -- Steven Spielberg

Arrogance Maxim: The ease of defeating a security device or system is proportional to how confident/arrogant the designer, manufacturer, or user is about it, and to how often they use words like "impossible" or "tamper-proof".

Shaw's Law: Build a system that even a fool can use, and only a fool will want to use it.

A common mistake that people make when trying to design something completely foolproof is to underestimate the ingenuity of complete fools.
  -- Douglas Adams (1952-2001)

It is impossible to design a system so perfect that no one needs to be good.
  -- T.S. Eliot (1888-1965)

Rohrbach's Maxim: No security device, system, or program will ever be used properly (the way it was designed) all the time.

Rohrbach Was An Optimist Maxim: No security device, system, or program will ever be used properly.

Nothing is eaten as hot as it is cooked.
  -- Otto Frisch (1904-1979)

In the too-good-to-be-true department, Argonne National Laboratory classified ad:
TO BE GIVEN AWAY.
TURTLE – This small painted turtle was found last January in my back yard upside down and half frozen. It is missing both paws on one side and thus cannot be returned to the wild. Contact ....

Why did the blonde take her new scarf back to the store? Because it was too tight.

38 - Locks, Seals, & Tamper Detection

Some Key Points:

• **All locks can be defeated, often quite easily.**

• **Don't put the information needed to check a tamper-indicating seal on the seal itself!**

• **Seals only reliably detect tampering if you have a good seal use protocol, which is the complete set of official and unofficial procedures for seal procurement, transport, storage, accountability, installation, inspection, removal, disposal, and training.**

There are no bad dogs, only inexperienced owners.
-- Barbara Woodhouse (1910-1988)

It is folly to bolt a door with a boiled carrot.
-- English proverb

Locks keep honest people honest.
-- old adage

A lock is meant only for honest men.
-- Yiddish proverb

Definition—TSA luggage lock: The TSA certifies that this device offers no meaningful security.

It is important to realize that any lock can be picked with a big enough hammer.
-- Sun System & Network Admin manual

I have six locks on my door all in a row. When I go out, I lock every other one. I figure no matter how long somebody stands there picking the locks, they are always locking three.
-- Elayne Boosler

If I had only known, I would have been a locksmith.

-- Albert Einstein (1879-1955)

-"Who are you and how did you get in here?"
-"I'm a locksmith. And, I'm a locksmith."
   -- Leslie Nielsen as Lieutenant Frank Drebin in *Police Squad*

For nature, heartless, witless nature,
Will neither care nor know
What stranger's feet may find the meadow
And trespass there and go.
Nor ask amid the dews of morning
If they are mine or no.
   -- Alfred Edward Housman (1859-1936)

A chicken is hatched even from such a well-sealed thing as an egg.
   -- Chinese proverb

Tamper Detection is a largely unsolved problem.
   -- Roger Johnston

The split tally was a kind of tag and seal used to establish a contract that would be hard to tamper with. The split tally or "tally sick" was used in medieval Europe and in rural areas of Germany and Switzerland into the 20th century, especially by illiterate people. A stick (often Hazelwood) was marked with circumferential notches and then split length-wise. Each party in the "contract" would keep half the stick. Any attempt to unilaterally add notches on one-half of the stick by one of the parties (thus tampering with the agreed upon contract) could be detected when the halves were reunited. The unique knots and surface markings on the stick would make counterfeiting challenging. The spit tally was a legally-binding contract, recognized by medieval courts and the 1804 Napoleonic Code.

Definition—Tamper-Indicating Seal = Seal: A magic security product that stops bad guys cold because they are afraid of the fact that it is called "tamper-proof".

"The time has come," the Walrus said,
"To talk of many things:
Of shoes—and ships—and sealing wax
Of cabbages—and kings."
   -- Lewis Carroll (1832-1898), *Through the Looking Glass*

There's no such thing as a "tamper-proof" or "tamper-resistant" seal. Those terms don't even make sense. Locks resist tampering, not seals. And the last thing we want is a seal that can't be tampered with. We want a seal that is easy for the bad guys to tamper with, but also easy for the good guys to tell that tampering has taken place.
 -- Roger Johnston

Definition—tamper-evident packaging: A strategy for reducing jury awards when product tampering inevitably occurs.

On a bag of Fritos: "You could be a winner! No purchase necessary. Details inside."

For manipulation to be most effective, evidence of its presence should be nonexistent.
 -- Herbert Schiller (1919-2000)

SEAL, n. A mark impressed upon certain kinds of documents to attest their authenticity and authority. Sometimes it is stamped upon wax, and attached to the paper, sometimes into the paper itself. Sealing, in this sense, is a survival of an ancient custom of inscribing important papers with cabalistic words or signs to give them a magical efficacy independent of the authority that they represent. In the British museum are preserved many ancient papers, mostly of a sacerdotal character, validated by necromantic pentagrams and other devices, frequently initial letters of words to conjure with; and in many instances these are attached in the same way that seals are appended now. As nearly every reasonless and apparently meaningless custom, rite or observance of modern times had origin in some remote utility, it is pleasing to note an example of ancient nonsense evolving in the process of ages into something really useful. Our word "sincere" is derived from sine cero, without wax, but the learned are not in agreement as to whether this refers to the absence of the cabalistic signs, or to that of the wax with which letters were formerly closed from public scrutiny. Either view of the matter will serve one in immediate need of an hypothesis. The initials L.S., commonly appended to signatures of legal documents, mean *locum sigillis*, the place of the seal, although the seal is no longer used—an admirable example of conservatism distinguishing Man from the beasts that perish. The words locum sigillis are humbly suggested as a suitable motto for the Pribyloff Islands whenever they shall take their place as a sovereign State of the American Union.
 -- Ambrose Bierce (1842-1914?), *The Devil's Dictionary*

Defeating seals is mostly about fooling people, not beating hardware.
-- Roger Johnston

Factoid: A seal is no better than its use protocol.

"Do not eat if seal is missing."
-- Printing on the seal for a consumer product

You know there's a problem when the people writing international standards for seals don't seem to fully grasp what a seal is.
-- Roger Johnston on ISO Standard 177712

A combination and a form indeed,
Where every god did seem to set his seal,
To give the world assurance of a man.
-- William Shakespeare (1564-1616), *Hamlet*, 3:4

Keep your valuables in a strong room,
let your wife not know what's in your purse.
Those who came before use established this,
our fathers only divided income with the gods.
They drove in a peg, made firm a ring, and attached clay (i.e., they sealed the door).
Keep your seal on a ring.
Surround the ring, and guard your house.
May your seals be the only access to your assets.
-- Approximately 4,000 year-old advice from a Babylonian father (Shupe-Ameli) to his son. From an inscription on clay tablets found throughout the Middle East. (Apparently the advice was popular and frequently reprinted.)

39 - Fakes, Frauds, Counterfeits, & Tags

Some Key Points:

• **Counterfeiting is much easier than most people think but mimicry is usually sufficient, and much easier.**

• **Current anti-counterfeiting tags aren't very good.**

I woke up the other morning and found that everything in my room had been replaced by an exact replica.
— Steven Wright

Manufacturers often claim to have a unique manufacturing process that nobody else can duplicate. In my experience, this is almost always exaggerated. But it doesn't really matter. The point often overlooked when considering the counterfeiting of a product or spoofing of a security device is that the adversary only needs to make the counterfeit superficially resemble the original, and maybe perform a few of its functions in roughly the same way. This is much easier than true counterfeiting.
-- Roger Johnston

Only God can tell the saintly from the suburban,
Counterfeit values always resemble the true;
Neither in Life nor Art is honesty bohemian,
The free behave much as the respectable do.
-- W.H. Auden (1907-1973)

Factoid: The great film comedian, Charlie Chaplin, once entered a Charlie Chaplin look-a-like contest for a laugh. To his surprise, he did not win.

Everyone wants to be Cary Grant. Even I want to be Cary Grant.
-- Cary Grant (1904-1986)

Actor David Niven, a colonel in the British army, recruited the actor E. Clifton James to impersonate British General Bernard Montgomery during World War II. James looked remarkably like Montgomery, and learned to mimic his speech, mannerisms, and bearing. The ruse was meant to fool the Germans as to where and when the D-Day invasion would take place. The story is told in James' 1955 book, *I was Monty's Double*.

If you can't imitate him, don't copy him.
    -- Yogi Berra (1925-2015)

"I am Spartacus."
    -- From Stanley Kubrick's movie, *Spartacus* (1960)

Man in Stadium Crowd: "Hey Look! It's Enrico Pallazzo!"
    -- From the movie, *The Naked Gun: From the Files of Police Squad* (1988)

The handwriting on the wall may be a forgery.
    -- Ralph Hodgson (1871-1962)

Our ability to manufacture fraud now exceeds our ability to detect it.
    -- From the movie, *S1mOne* (2002)

Factoid: It is estimated that only 1% of "Louis Vuitton" designer purses are authentic.

Boycott shampoo! Demand the REAL poo!
    -- Steven Wright

Definition—product anti-counterfeiting tag: Something a manufacturer or product counterfeiter sticks on a product to make the customer think it is authentic.

It makes no sense to include a pamphlet along with your product telling customers how to spot counterfeits, because product counterfeiters could do the same thing. There has to be a second channel of trusted communication. This is obvious, but apparently not to a lot of manufacturers.
    -- Roger Johnston

Factoid: a remarkably high percentage of art and museum artifacts on display in museums worldwide are fake.

Factoid: For every real flamingo in the United States there are 700 plastic ones.

Inspector Gadget: Be careful, Brain, those are probably priceless fake artifacts.
    -- From the movie, *Inspector Gadget* (1983)

Fakes

You can't fake quality any more than you can fake a good meal.
    -- William Seward Burroughs (1914-1997)

Nothing is as apt to sharpen one's ability to discern the genuine as the recognition of the false.
    -- Friedrich Winkler (1888-1965)

Joe Banks: I've never been to L.A. before.
Angelica: What do you think?
Joe Banks: It looks fake. I like it!
    -- Dialog from the (unfairly panned) movie, *Joe Versus the Volcano* (1990)

There are really 4 different kinds of tags, depending on whether counterfeiting and/or lifting are of concern. ("Lifting" is removing the tag and sticking it on something else without being detected.) People talking about tags are usually vague on which kind they have in mind. The 4 kinds are...
• inventory tag (neither lifting nor counterfeiting is of concern)
• security tag (both lifting and counterfeiting are of concern)
• anti-counterfeiting tag (only counterfeiting is of concern)
• buddy tag or token (only counterfeiting is of concern)

Tag, you're it.
    -- children's game

Nothing is like it seems, but everything is exactly like it is.
    -- Yogi Berra (1925-2015)

Everything is what it is, and not another thing.
    -- Bishop Joseph Butler (1692-1752)

40 - Biometrics & Access Control

Some Key Points:

• **Biometrics—techniques for identifying people or verifying their identity based on their unique biological characteristics—are not an automatic security panacea.**

• **Most biometric and access control devices lack effective tamper detection (and sometimes have none), making them easy to spoof. (Counterfeiting biometric signatures is not all that difficult, either.)**

• **Access control is harder than it looks.**

I do not care to belong to a club that accepts someone like me as a member.
  -- Groucho Marx (1890-1977)

I was the kid next door's imaginary friend.
  -- Emo Philips

God has given you one face, and you make yourself another.
  -- William Shakespeare (1564-1616)

If I were two-faced, would I be wearing this one?
  -- Abraham Lincoln (1809-1865)

I never forget a face, but I'll make an exception in your case.
  -- Groucho Marx (1890-1977)

I'm always amazed to hear of accident victims being identified by their dental records. If they don't know who you are, how do they know who your dentist is?
  -- Paul Merton

I'm holding out for *trimetrics*.
  -- Anonymous

As of tomorrow, employees will only be able to access the building using individual security cards. Pictures will be taken next Wednesday and employees will receive their cards in two weeks.
    -- Email at Microsoft

"Badges? We don't need no stinkin' badges!"
    -- From the movie, *The Treasure of the Sierra Madre* (1948) [The actual dialog was, "Badges? We ain't got no badges. We don't need no badges! I don't have to show you any stinkin' badges!"]

Factoid: Tongue prints and ear morphology are unique to each individual. Ear shapes were used in the 1880's to identify criminals.

Definition—keyless entry: If you poke it with a paperclip or a credit card, the door will open.

You can go anywhere if you look serious and carry a clipboard.
    -- Scott Adams (*Dilbert*)

The easiest way to get into a limited access area is to be an attractive young woman, and have your hands full with coffee and papers.
    -- Anonymous

Factoid: A mechanical tamper switch is functionally about the same as having no tamper detection at all.

Police Officer: Could I have your name?
Kip Wilson: Well, you could, but it would be an incredible coincidence.
    -- *Bosom Buddies*

The first time we ever made love, I said, "Am I the first man who ever made love to you?" She said, "You could be. You look damn familiar."
    -- Ronnie Bullard

Actual 911 recorded call:
911: 911. What is the nature of your emergency?
Caller: My wife is pregnant, and her contractions are only two minutes apart!
911: Is this her first child?
Caller: No, you idiot! This is her husband!

"We've been searching for you, God!"

-- Computer-generated junk mail from American Family Publishers to the Bushnell Assembly of God Church

41 - Alcohol, Drugs, & Drug Testing

Some Key Points:

• **Drugs & alcohol are pervasive.**

• **Given that the results of drug testing have important implications for national security, safety, sports, employers, and an employee's livelihood, career, and reputation, better drug testing security would seem to be warranted.**

Looks like I picked the wrong week to stop sniffing glue.
  -- From the movie, *Airplane!* (1980)

I hate to advocate drugs, alcohol, violence, or insanity to anyone, but they've always worked for me.
  -- Hunter S. Thompson (1937-2005)

"Warning: do not use if you have prostate problems."
  -- On a box of Midol PMS relief tablets

Factoid: The word "drunk" is believed to have the most number of synonyms of any English word: 2,231.

Factoid: About 1 million Americans are arrested for drunkenness each year.

Bacchus hath drowned more men than Neptune.
  -- Thomas Fuller (1654-1734)

I always keep a supply of stimulant handy in case I see a snake—which I also keep handy.
  -- W.C. Fields (1880-1946)

If the doctor always says to take two aspirins, why don't they just sell them twice as big?
  -- Anonymous

Printed on a Chinese medicine bottle: "Expiration date: 2 years."

Why do drug stores make sick people walk all the way to the back of the store to get prescriptions, while healthy people can buy cigarettes up front?
-- Anonymous

Factoid: The probability of getting a cold after a 2-hour plane flight: 20%

Definition—Drug: A substance, which when injected into a rat, produces a scientific report.
-- Anonymous

Q: Why can't blondes be pharmacists?
A: It is too hard to get vials and bottles into the typewriter.

To keep my job, I have to pee into a cup every once in a while. If they showed as much interest in me and my work as they do in my urine, this would be a good place to work and we'd have good security.
-- Anonymous

It's just too bad that urine testing can't screen for stupidity.
-- Anonymous

Beauty is in the eye of the beerholder.
-- W.C. Fields (1880-1946)

I dunno. I never smoked any Astroturf.
-- Baseball player Tug McGraw, asked whether he preferred grass or Astroturf

Now, like, I'm president. It would be pretty hard for some drug guy to come into the White House and start offering it up, you know? I bet if they did, I hope I would say, "Hey, get lost! We don't want any of that."
-- President George W. Bush talking to students about drug abuse

After the first glass you see things as you wish they were. After the second, you see things as they are not. Finally, you see things as they really are, and that is the most horrible thing in the world!
-- Oscar Wilde (1854-1900), on consumption of Absinthe

I'm not fat.
-- Boxer Randall "Tex" Cobb after a reporter called him a fat, cocaine-snorting drunk

I can honestly say, all the bad things that ever happened to me were directly, directly attributed to drugs and alcohol. I mean, I would never urinate at the Alamo at nine o'clock in the morning dressed in a woman's evening dress sober.
-- Ozzy Osbourne

Based on our analysis, the security of urine testing products is very poor. This is unfortunate given that a lot is riding on the results for both an employee and the employer.
-- Roger Johnston

42 - Maybe Missing the Point?

Some Key Points:

• **It's easy to miss the forest for the trees, the questions for the answers, and the causes for the effects.**

It isn't that they can't see the solution.
It is that they can't see the problem.
    -- G.K. Chesterton, *The Scandal of Father Brown* (1935)

"How Green Were the Nazis?"
    -- Actual book title

Never answer an anonymous letter.
    -- Yogi Berra (1925-2015)

CLARIFICATION: It has come to the editor's attention that the *Herald-Leader* neglected to cover the civil rights movement. We regret the omission.
    -- Lexington (KY) *Herald-Leader*, July 4, 2004

It's really hard to maintain a one-on-one relationship if the other person is not going to allow me to be with other people.
    -- Axl Rose

I don't care what is written about me as long as it isn't true.
    -- Katherine Hepburn (1907-2003)

Sign outside a card shop: "I Love You Only" Valentine's Day cards. Now available in multi-packs!

Q: What would you like most for Christmas? (Asked in 1948 by a Washington, DC radio station of different ambassadors.)
A: French Ambassador--Peace throughout the world.
A: Soviet Ambassador--Freedom for all people enslaved by imperialism.
A: British Ambassador, Sir Oliver Franks--Well, it's very kind of you to ask. I'd quite like a box of crystallized fruit.

43 - Nuclear Nonproliferation & Safeguards

Some Key Points:

• **Even a security application as important and serious as nuclear safeguards isn't necessarily well thought through.**

Arms control has to have a future, or none of us does. But it doesn't necessarily have to come in big packages of 600-page treaties.
-- Stanley Hoffmann

Instead of building newer and larger weapons of mass destruction, I think mankind should try to get more use out of the ones we have.
-- Jack Handey

You can't be a Real Country unless you have a BEER and an airline. It helps if you have some kind of a football team, or some nuclear weapons, but at the very least you need a BEER.
-- Frank Zappa (1940-1993)

One nuclear explosion can ruin your whole day.
-- Anonymous

War does not determine who is right—only who is left.
-- Bertrand Russell (1872-1970)

The best defense against the atom bomb is not to be there when it goes off.
-- British Army Journal, 1949

All you have to do is go down to the bottom of your swimming pool and hold your breath.
-- DOE spokesman, on protecting yourself from nuclear radiation

I do not like this word "bomb". It is not a bomb. It is a device that is exploding.
-- French Ambassador Jacques le Blanc on nuclear weapons

A lot of safeguards and radiological monitoring equipment lacks effective tamper detection. Sometimes there is none at all.
-- Roger Johnston

Nuclear Material Control and Accounting (MC&A) looks superficially like an inventory function, but it is really a security function. If the supposed MC&A system makes no effort to counter spoofing (as is often the case), it has no role to play in security, and it can provide little meaningful information about theft or diversion.
-- Roger Johnston

Gunslingers' Maxim: Any government security program will mistakenly focus more on dealing with force-on-force attacks than on the more likely attacks involving insider threats and more subtle, surreptitious attacks.

Be sure and put some of those neutrons on it.
-- Baseball pitcher Mike Smith, ordering a salad at a restaurant

I don't know what's scarier, losing nuclear weapons, or that it happens so often there's actually a term for it.
-- From the movie, *Broken Arrow* (1996)

The local council at Barnsley has had to abandon a civil defense post designed to survive a nuclear war because it has been wrecked by vandals.
-- *The Daily Telegraph*

Then there was the time that my garage became a licensed radiological storage site. We had this cat that was treated for feline hyper-thyroidism with radioactive iodine. Because the county dump in Los Alamos, New Mexico was one of the few municipal landfills in the country with radiological sensors (due to Los Alamos National Laboratory being nearby), the radioactive kitty litter would have set off alarms. So we had to store the kitty litter in my garage—after proper licensing—for several months until enough half-lives had passed for it to be taken to the dump. That cat was a pain in the butt for other reasons, too.
-- Roger Johnston

44 - Terminology

Some Key Points:

• A rose is a rose by any other name, but bad (or hijacked) terminology causes sloppy thinking and conceptual errors.

• Stay alert to how language can constrain, distort, or mislead your reasoning.

Lisa, Bart, and Homer Simpson:
 -A rose by any other name would smell as sweet.
 -Not if you called 'em stenchblossoms.
 -Or crapweeds

I once had a rose named after me and I was very flattered. But I was not pleased to read the description in the catalogue: no good in a bed, but fine up against a wall.
   -- Eleanor Roosevelt (1884-1962)

All we know is, when you put these two words together, it's magic.
   -- Scott MacHardy, on his company's popular "Coed Naked" clothing line

People often confuse "vulnerabilities" (security weaknesses that a threat could exploit) with threats, assets needing protection, attack scenarios, or features of their facility or security. Similarly, "vulnerability assessments" get confused all the time with analysis methods that aren't nearly as good at finding vulnerabilities: security surveys, pen testing, fault/event trees, DBT, "Red Teaming", risk assessment, etc. When security terms get confused or even hijacked, it becomes difficult to speak and think carefully about security issues.
   -- Roger Johnston

How many legs does a dog have if you call the tail a leg? Four. Calling a tail a leg doesn't make it a leg.
   -- Abraham Lincoln (1809-1865)

Nothing sucks like an Electrolux!
   -- Original advertising slogan for Electrolux vacuums, written by Swedes who
   didn't understand the English language nuances of the work "suck"

The slovenliness of our language makes it easier for us to have foolish thoughts.
-- George Orwell (1903-1950)

The great enemy of clear language is insincerity.
-- George Orwell (1903-1950)

When words lose their meaning, there is chaos in the land.
-- Confucius (551 BC – 479 BC)

Factoid: The first James Bond Movie, *Dr. No*, was released in Japan under the title, *We Have No Need of a Doctor.*

A tourist once stopped to admire a North Carolina mule. He asked the mule's owner what the animal's name was. The farmer said, "I don't know, but we call it Bill"
-- Samuel James Ervin, Jr.

In real life, unlike in Shakespeare, the sweetness of the rose depends upon the name it bears. Things are not only what they are, they are, in a very important respect, what they seem to be.
-- Hubert H. Humphrey (1911-1978)

Actual courtroom testimony:
Q: What was the first thing your husband said to you when he woke up that morning?
A: He said, "Where am I, Cathy?"
Q: And why did that upset you?
A: My name is Susan.

Definition—security: The absence of anything bad happening due to dumb luck.

Definition—best practice: Those guys don't know what the hell they are doing either, but at least they seem confident.

Why is it that when you transport something by car, it's called a shipment, but when you transport something by ship, it's called cargo?
-- Anonymous

Factoid: "Jumping the Shark" is a term used when a television series is running out of ideas and desperately trying to hang onto its audience. The term originated from an episode of the TV show *Happy Days* (1974-1984) when Fonzie jumped over sharks on waterskis.

I believe the best definition of man is the ungrateful biped.
  -- Feodor Dostoevsky (1821-1881)

I'd say, "It's a Buttmaster, Your Holiness."
  -- Suzanne Somers on how she would respond if the Pope asked her the name of the exercise machine she promotes

After the police break into Mimi's dressing room:
Mimi du Jour: Is this some kind of bust?
Frank Drebin: Yes, ma'am, it's very impressive, but we need to ask you a few questions.
  -- From the television show, *Police Squad* (1982)

A cynic is a person who tells the truth, but probably could've phrased it better.
  -- Roger Johnston

Scratch any cynic, and you'll find a disappointed idealist.
  -- George Carlin (1937-2008)

No matter how cynical you become, it is never enough to keep up.
  -- Lily Tomlin

Why are a wise man and a wise guy opposites?
  -- Anonymous

Definition—Dooking: The sounds ferrets make when happy.

The problem with the French, is that they don't have a word for 'entrepreneur'.
  -- George W. Bush

"Ich bin ein Berliner." [I am a jelly donut.]
  -- John F. Kennedy (1917-1963)

Who's the joker who put the letter 's' in the word 'lisp'?
  -- Anonymous

You 'whisper' the air out of the container...We don't say 'burp' anymore. It's rude.
  -- Colleen Staab, Tupperware Sales Manager

I don't even know whether "hip" is the word for hip anymore.
-- Anonymous

Too hell with the cost! If it's a good story, I'll make it.
-- Movie mogul Samuel Goldwyn when told a script was "too caustic"

That cartoon character, Asterix. I wonder how rude his real name is.
-- Jimmy Carr

The Democrats are going to change the name of the Hoover Dam. That is the silliest thing I ever heard of in politics…If they feel that way about it, I don't see why they don't just reverse the two words.
-- Will Rogers (1879-1935)

A tourist once stopped to admire a North Carolina mule. He asked the mule's owner what the animal's name was. The farmer said, "I don't know, but we call it Bill."
-- Samuel James Ervin, Jr.

The word is not covert, it's overt. Covert means you're out in the open. Overt is what I did. I was undercover.
-- Nevada Senator Chic Hecht commenting on his 18 years of covert operations as an Army intelligence   officer

Seal users often assume (incorrectly) that the tamper-indicating seals they are using cannot be spoofed because the vendor or manufacturer refers to them as "tamper-proof seals".
-- Roger Johnston

Factoid: The Patagonian Toothfish (*Dissostichus eleginoides*) was not a popular fish for dining until it was renamed "Chilean Sea Bass". Similarly, Orange Roughy (*Hoplostethus atlanticus*) didn't sell too well when it was called "Slimehead".

Chuck Berry penned the words, "My ding-a-ling, my ding-a-ling, I want you to play with my ding-a-ling." If only he'd given a thought to how those lyrics could be construed by lewd-minded folks.
-- Mrs. Merton

I have been called dumb, crazy man, science abuser, Holocaust denier, villain of the month, hate-filled, warmonger, Neanderthal, Genghis Khan, and Attila the Hun. And I can just tell you that I wear some of those titles proudly.

Terminology

-- Oklahoma Senator James Inhofe

45 - Security & Change

Some Key Points:

- **Change is inevitable.**

- **Change will usually be feared and resisted.**

- **Good security requires flexibility and constant change.**

People generally prefer the predictable. Few recognize how
destructive this can be, how it imposes severe limits on variability
and thus makes whole populations fatally vulnerable to the
shocking ways our universe can throw the dice.

> -- Frank Herbert (1920-1986)

Don't give a permanent solution to a temporary situation.
> -- Martin Uzochukwu Ugwu

The dinosaurs became extinct because they didn't have a space program.
> -- Larry Niven

It is not the strongest of the species that survive, not the most intelligent, but the
one most responsive to change.
> -- Charles Darwin (1809-1882)

New ideas pass through three periods: It can't be done. It probably can be done
but it's not worth it. I knew it was a good idea all along.
> -- Arthur C. Clarke (1917-2008)

When a thing is new, people say: "It is not true." Later, when its truth becomes
obvious, they say: "It's not important."   Finally, when its importance cannot be
denied, they say, "Anyway, it's not new."
> -- William James (1842-1910)

All truth passes through three stages. First, it is ridiculed. Second, it is violently opposed. Third, it is accepted as being self-evident.
-- Arthur Schopenhauer (1788-1860)

First they ignore you, then they laugh at you, then they fight you, then you win.
-- Mahatma Ghandi (1869-1948)

All great truths begin as blasphemies.
-- George Bernard Shaw (1856-1950)

Whenever I hear, 'It can't be done,' I know I'm close to success.
-- Michael Flatley

It's the most unhappy people who most fear change.
-- Mignon McLaughlin (1913-1983)

The only thing worse than change simply for the sake of change is stagnation simply for the sake of comfort.   -- Anonymous

Even cowards can endure hardship; only the brave can endure suspense.
-- Mignon McLaughlin (1913-1983)

I can't understand why people are frightened of new ideas. I'm frightened of the old ones.
-- John Cage (1912-1992)

No matter how brilliantly an idea is stated, we will not really be moved unless we have already half thought of it ourselves.
-- Mignon McLaughlin (1913-1983)

To improve is to change; to be perfect is to change often.
-- Winston Churchill (1874-1965)

To try to be better is to be better.
-- Charlotte Saunders Cushman (1816-1876)

Without deviation from the norm, progress is not possible.
-- Frank Zappa (1940-1993)

If we continue to do things the way they have always been done, the most we can expect is what we already have.
-- Dennis Bay

If you always do what you always did, you will always get what you always got.
-- Moms Mabley (1894-1975)

When you come to a fork in the road, take it.
-- Yogi Berra (1925-2015)

Change means the unknown... It means too many people cry, insecurity. Nonsense! No one from the beginning of time has had security.
-- Eleanor Roosevelt (1884-1962)

Only in growth, reform, and change, paradoxically enough, is true security to be found.
-- Anne Morrow Lindbergh (1906-2001)

Security can only be achieved through constant change, through discarding old ideas that have outlived their usefulness and adapting others to current facts.
-- William O. Douglas (1898-1980)

Our only security is our ability to change.
-- John Lilly (1915-2001)

Security . . . it's simply the recognition that changes will take place and the knowledge that you're willing to deal with whatever happens.
-- Harry Browne (1933-2006)

The search for static security—in the law and elsewhere—is misguided. The fact is security can only be achieved through constant change, adapting old ideas that have outlived their usefulness to current facts.
-- A. William Osler

Even if you're on the right track, you'll get run over if you just sit there.
-- Will Rogers (1879-1935)

It takes a lot of courage to release the familiar and seemingly secure, to embrace the new. But there is no real security in what is no longer meaningful. There is more security in the adventurous and exciting, for in movement there is life, and in change there is power.

-- Alan Cohen

Most advances in science come when a person is forced to change fields.
    -- Peter Borden, physicist

'Tis Providence alone secures
In every change both mine and yours.
    -- Cowper, A Fable: Moral

The great thing about television is that if something important happens anywhere in the world, day or night, you can always change the channel.
    -- Rev. Jim Ignatowski, *Taxi*

Eternity magazine will cease publishing with the January issue.
    -- Press release

In a world in which the total of human knowledge is doubling about every ten years, our security can rest only on our ability to learn.
    -- Nathaniel Branden

Don't worry about people stealing an idea. If it's original, you will have to ram it down their throats.
    -- Howard Aiken (1900-1973)

Progression is not proclamation nor palaver. It is not pretense nor play on prejudice...It is not the perturbation of a people passion-wrought, nor a promise proposed.
    -- Warren G. Harding (1865-1923)

Headline in the Financial Times: BAKED BEAN RIVALS BEGIN TO FEEL THE WIND OF CHANGE

46 - Technology

Some Key Points:

• **High-Tech is not a silver bullet.**

• **Technology is a useful tool for security, not an excuse to stop thinking critically.**

A healthy sense of all we don't know-even a sense of mystery-keeps us from reaching for oversimplifications and technological silver bullets.
  -- Michael Pollan

Technology can relieve the symptoms of a problem without affecting the underlying causes. Faith in technology as the ultimate solution to all problems can thus divert our attention from the most fundamental problem—the problem of growth in a finite system—and prevent us from taking effective action to solve it.
  -- Donella A. Meadows

High tech is potent, precise, and in the end, unbeatable. The truth is, it reminds a lot of people of the way I pitch horseshoes. (Would you believe some of the people? Would you believe our dog?) Look, I want to give the high-five symbol to high tech.
  -- George H.W. Bush

Inanimate objects can be classified scientifically into three major categories: those that don't work, those that break down, and those that get lost.
  -- Russell Baker

When I win the lottery, I'm going to fund a Wile E. Coyote Chair of Applied Engineering at some University.
  -- William Starr

"Not dishwasher safe."
  -- On a remote control for a TV

Technology: No place for wimps!
  -- Scott Adams (*Dilbert*)

Definition—high-tech: (1) The owner's manual is badly written. (2) An excuse to stop thinking about security.

Definition—next generation: Our components are so old, we can't buy them anymore.

When the only tool you have is a hammer, all problems begin to resemble nails.
-- Abraham Maslow (1908-1970)

If we were to go back in time 100 years and ask a farmer what he'd like if he could have anything, he'd probably say he wanted a horse that was twice as strong and ate half as many oats. He would not say he wanted a tractor. The point is, technology changes things so fast that many people aren't sure what the best solutions to their problems might be.
-- Philip Quigley

Few of the mainframe computer users that IBM surveyed in the 1970's could imagine why they would ever want a small computer on their desk.
-- Robert I. Sutton

The marvels of modern technology include the development of a soda can which, when discarded, will last forever, and a $17,000 car which, when properly cared for, will rust out in two or three years.
-- Paul Harwitz

There is a proper tool to misuse for every task.
-- Anonymous

Byrne's Law: In any electrical circuit, appliances and wiring will burn out to protect the fuses.

All technology expands the space, contracts the time, and destroys the working group.
-- Vint Cerf

Why don't they make the whole plane out of that black box stuff?
-- Steven Wright.

Buffett's Maxim: You should only use security hardware, software, and strategies you understand.

Don't deploy any security technology until you have at least some modest understanding of how it works. Once you obtain this understanding, you may well come to realize that it's a can of worms, not a silver bullet.
    -- Roger Johnston

Duct tape is like the Force. It has a light side, a dark side, and it holds the universe together.
    -- Carl Zwan

Factoid: A cricket chirp can be heard at a distance of a mile under good conditions.

The badness of a movie is directly proportional to the number of helicopters in it.
    -- Dave Barry

Heisenberg's cat pissed on my quantum computer!
    -- Bumper Sticker

Engineers don't understand security. They have a different mindset than the bad guys.
    -- Roger Johnston

For 'Tis the sport to have the engineer hoisted with his own petard.
    -- William Shakespeare (1564-1616), Hamlet, 3:4

Technology frightens me to death. It's designed by engineers to impress other engineers, and they always come with instruction booklets that are written by engineers for other engineers—which is why almost no technology ever works.
    -- John Cleese

Technology... is a queer thing. It brings you great gifts with one hand, and it stabs you in the back with the other.
    -- Carrie P. Snow

We live in a Newtonian world of Einsteinian physics ruled by Frankenstein logic.
    -- David Russell

The march of science and technology does not imply growing intellectual complexity in the lives of most people. It often means the opposite.
    -- Thomas Sowel

Why doesn't Tarzan have a beard?

-- Anonymous

As a scientist at a national lab, I was always amazed at government agencies visiting the laboratory shopping for technology. They have very specific criteria: (1) The technology has to be a breakthrough like a tricorder, invisibility cloak, or time machine. (2) No additional R&D is needed to field it. (3) The technology must be inexpensive, can be run by morons in the field, and only gives a red light – green light response. (4) It has been sitting on the shelf not being used by anybody else because—while we geniuses invented the tricorder, invisibility cloak, or time machine—we are too stupid to think of a use for it.
   -- Roger Johnston

Where facts are few, experts are many.
   -- Donald R. Gannon

I have a friend who failed the Turing Test.
   -- Rob Munsch

How do you know when you've run out of invisible ink?
   -- Steven Wright

In contemplating the use of any new security measure, strategy, or product, you need to determine the correct answers to 3 questions:
   (1) To what extend does this really improve security?
   (2) What are all the costs, tradeoffs, and side effects (because there always are some)?
   (3) Is 1 commensurate with 2?
     -- Roger Johnston

The scientific theory I like best is that the rings of Saturn are composed entirely of lost airline luggage.
   -- Mark Russell

How do they get those dead bugs into those enclosed light fixtures?
   -- Anonymous

The wireless telegraph is not difficult to understand. The ordinary telegraph is like a very long cat. You pull the tail in New York, and it meows in Los Angeles. The wireless is the same, only without the cat.
   -- Albert Einstein (1879-1955)

Probably the earliest flyswatters were nothing more than some sort of striking surface attached to the end of a long stick.
 -- Jack Handey

Putt's Law: Technology is dominated by two types of people: those who understand what they do not manage, and those who manage what they do not understand.

I was walking down the street wearing glasses when the prescription ran out.
 -- Steven Wright

Factoid: Diamonds burn, turning into pure carbon dioxide gas. The minerals lonsdaleite and wurtzite boron nitride in pure form are 58% and 18% harder than diamonds, respectively.

The three fundamental Rules of Robotics:
1. A robot may not injure a human being, or through inaction allow a human being to come to harm.
2. A robot must obey the orders given it by human beings except where such orders would conflict with the First Law.
3. A robot must protect its own existence as long as such protection does not conflict with the First and Second Laws.
 -- Isaac Asimov (1920-1992)

In real life, I assure you, there is no such thing as algebra.
 -- Fran Lebowitz

I still have a hard time believing that toilets work without electricity.
 -- John Maclain

I have always wished that my computer would be as easy to use as my telephone. My wish has come true. I no longer know how to use my telephone.
 -- Bjarne Stroustrup

If you think technology can solve your security problems, then you don't understand the problems and you don't understand the technology.
 -- Bruce Schneier

All technology should be assumed guilty until proven innocent.
 -- David Brower (1912-2000)

Play the music, not the instrument.
>   -- Anonymous

For a successful technology, reality must take precedence over public relations, for Nature cannot be fooled.
>   -- Richard Feynman (1918-1988)

There are two kinds of fool. One says, "This is old, and therefore good." And one says, "This is new, and therefore better."
>   -- John Brunner (1934-1995)

Any sufficiently advanced technology is indistinguishable from a rigged demo.
>   -- Arthur C. Clarke (1917-2008)

The more sophisticated the technology, the more vulnerable it is to primitive attack. People often overlook the obvious.
>   -- Tom Baker as Doctor Who in *The Pirate Planet* (1978)

Technology... the knack of so arranging the world that we don't have to experience it.
>   -- Max Frisch (1911-1991)

The Air Force is pleased with the performance of the C-5A cargo plane, although having the wings fall off at eight thousand hours is a problem.
>   -- Major General Charles F. Kyuk, Jr.

The guy who invented the wheel was an idiot. The guy who invented the other three, he was a genius.
>   -- Sid Caesar (1922-2014)

New technology is common, new thinking is rare.
>   -- Sir Peter Blake (1948-2001)

47 - The Future

Some Key Points:

• **Nobody predicts the future well, especially experts.**

• **So stay flexible!**

• **The past often isn't a good basis to predict the future (or future security incidents).**

• **Good ideas have a freshness date.**

Premier Zhou Enlai, when asked about the effect of the French Revolution, said it was too early to tell. (Actually, the quote turns out to be a misunderstanding. Zhou thought he was being asked about the 1968 French student street riots.)

Firestone's Law: Chicken Little only has to be right once.

Prediction is very difficult, especially about the future.
    -- Niels Bohr (1885-1962)

It's tough to make predictions, especially about the future.
    -- Yogi Berra (1925-2015)

I never predict anything, and I never will.
    -- English soccer player Paul John Gascoigne

If you can look into the seeds of time, and say which grain will grow and which will not, speak then unto me.
    -- William Shakespeare (1564-1616), Macbeth, 1:3

We drive into the future using only our rearview mirror.
    -- Marshall McLuhan (1911-1980)

You can never plan the future by the past.
    -- Edmund Burke (1729-1797)

The Future

History is more or less bunk.
-- Henry Ford (1863-1947)

A good forecaster is not smarter than everyone else, he merely has his ignorance better organized.
-- Anonymous

The future ain't what it used to be.
-- Yogi Berra (1925-2015)

Remember, today is the tomorrow you worried about yesterday.
-- Dale Carnegie (1888-1955)

"You may have already won."
-- Common contest come-on

Things are more like they are now than they ever were before.
-- Dwight D. Eisenhower (1890-1969)

Things are more like they are now than they have ever been.
-- Gerald Ford (1913-2006)

The future will be better tomorrow.
-- Dan Quayle

The future is assured. It's just the past that keeps changing.
-- Traditional Russian joke

If it's the "Psychic Network" why do they need a phone number?
-- Steve Martin

The learned usually find themselves equipped to live in a world that no longer exists.
-- Eric Hoffer (1902-1983)

I decided not to let my past rule my future so I decided to change my present in order to open up my future.
-- Ana M. Guzman

The empires of the future are the empires of the mind.
-- Winston Churchill (1874-1965)

The best way to predict the future is to invent it.
-- Alan Kay

The future is here. It's just not widely distributed yet.
-- William Gibson

I look to the future because that's where I'm going to spend the rest of my life.
-- George Burns (1896-1996)

The enemies of the future are always the very nicest people.
-- Christopher Morley (1890-1957)

Clarke's First Law: When a distinguished but elderly scientist states that something is possible, he is almost certainly right. When he states that something is impossible, he is very probably wrong.

Drill for oil? You mean drill into the ground to try and find oil? You're crazy.
-- Drillers whom Edwin L Drake tried to enlist to his project to drill for oil, 1859

Louis Pasteur's theory of germs is ridiculous fiction.
-- Pierre Pachet, Professor of Physiology at Toulouse, 1872

Law will be simplified over the next century. Lawyers will have diminished, and their fees will have been vastly curtailed.
-- Journalist Julius Henri Browne, 1893

Heavier than air flying machines are impossible.
-- Lord Kelvin, President, Royal Society, 1895

This 'telephone' has too many shortcomings to be seriously considered as a means of communication. The device is inherently of no value to us.
-- Western Union memo, 1876

I'm sorry Mr. Kipling, but you don't know how to use the English language.
-- Editor of the *San Francisco Examiner*, rejecting a short story from Rudyard Kipling

Everything that can be invented has been invented.
-- Charles H. Duell, Director of the U.S. Patent Office, 1899

Airplanes are interesting toys but of no military value.
    -- Marechal Ferdinand Foch, Professor of Strategy, Ecole Superieure de Guerre

The wireless music box has no imaginable commercial value. Who would pay for a message sent to nobody in particular?
    -- David Sarnoff's associates in response to his urgings for investment in the radio in the 1920's.

Who the hell wants to hear actors talk?
    -- H.M. Warner, Warner Bros, 1927

Fred Astaire can't act, can't sing, balding... Can dance a little.
    -- MGM talent scout, 1928

Stocks have reached what looks like a permanently high plateau.
    -- Irving Fisher, Economics Professor, Yale University, 1929

I think there's a world market for maybe five computers.
    -- Thomas Watson, chairman of IBM, 1943

There is no reason why anyone would want to have a computer in their home.
    -- Ken Olson, president, chairman and founder of Digital Equipment Corp, 1977

Computers in the future will weigh no more than 1.5 tons.
    -- Popular Mechanics, 1949

You ain't goin' nowhere son. You ought to go back to drivin' a truck.
    -- Jim Denny of the Grand Ole Opry, Nashville, firing Elvis Presley after his first performance

I have traveled the length and breadth of this country and talked with the best people, and I can assure you that data processing is a fad that won't last out the year.
    -- Editor in charge of business books for Prentice Hall, 1957

We don't like their sound, and guitar music is on the way out.
    -- Decca Recording Company rejecting the Beatles, 1962

A cookie store is a bad idea. Besides, the market research reports say that America likes crispy cookies, not soft and chewy cookies like you make.
    -- Response to Debbi Fields' idea of starting the Mrs. Fields Cookies business

The concept is interesting and well-formed, but in order to earn better than a 'C,' the idea must be feasible.
-- A Yale University management professor in response to Fred Smith's paper proposing overnight delivery service. Smith eventually founded Federal Express.

640K ought to be enough for anybody.
-- Bill Gates, 1981

All good ideas are eventually bad ideas.
-- Wall Street adage

All movements go too far.
-- Bertrand Russell (1872-1970)

It is defeats that make one a human being. A man who never understands his defeats takes nothing with him into the future.
-- Aksel Sandemose (1899-1965)

If you think you understand the way time is expressed in a novel, consider the sentence, "Tomorrow was Christmas."
-- Gordon Eugene Nelson

When you're riding in a time machine way far into the future, don't stick your elbow out the window, or it'll turn into a fossil.
-- Jack Handey

Just as we outgrow a pair of trousers, we outgrow acquaintances, libraries, principles, etc., at times before they're worn out and times—and this is the worst of all—before we have new ones.
-- Georg C. Lichtenberg (1742-1799)

You can count the seeds in an apple, but you can't count the apples in a seed.
-- Ken Kesey (1935-2001

About The Author

Roger G. Johnston, Ph.D., CPP has over 30 years of experience as a vulnerability assessor.

Books By This Author

Security Sound Bites: Important Ideas About Security From Smart-Ass, Dumb-Ass, and Kick-Ass Quotations

"Fascinating ... full of thought triggers"
-- *Security* Magazine

Vulnerability Assessment: The Missing Manual for the Missing Link
https://www.amazon.com/dp/B08C9D73Z9

You can't prevent or test what you have not envisioned!

Security usually fails because vulnerabilities and attack scenarios were not envisioned. This is often the weak link in the chain of security. A Vulnerability Assessment (VA) can help to fix the problem, but VAs are often missing or else get confused with other kinds of assessments and security "testing" that are not VAs, and are not very good at finding vulnerabilities, attack scenarios, or possible counter-measures. This book is the missing, comprehensive guide for how to actually do imaginative, quality VAs and find security problems. Along the way, tips for better security are offered.

Devil's Dictionary of Security Terms
https://www.amazon.com/dp/B08CP92PCC

Why go through your entire security, law enforcement, or intelligence career being confused? Here, at last, is an 850+ word dictionary to clarify all that confusing security jargon, by giving you the TRUE meaning of various terms, never mind what the experts think! Ideal reading material while sitting on the can!

www.ingramcontent.com/pod-product-compliance
Lightning Source LLC
LaVergne TN
LVHW042333060326
832902LV00006B/150